3/97

The Almanac of
Fascinating Beginnings

BOOKS BY NORMAN KING

Everybody Loves Oprah
Arsenio Hall
Madonna: The Book
The First Five Minutes
The Last Five Minutes
Donahue: The Man Women Love
Effective Advertising in Small Space
The Money Messiahs
Turn Your House Into a Money Factory
The Money Market Book
Two Royal Women
The Prince and the Princess: A Love Story
Here's Erma: The Bombecking of America
Dan Rather
Ivana
All in the First Family (with Bill Adler)
Hillary: Her True Story

The Almanac of Fascinating Beginnings

From the Academy Awards to the Xerox Machine

NORMAN KING

A Citadel Press Book
Published by Carol Publishing Group

A Citadel Press Book
Published by Carol Publishing Group
Citadel Press is a registered trademark of Carol Communications, Inc.
Editorial Offices: 600 Madison Avenue, New York, N.Y. 10022
Sales and Distribution Offices: 120 Enterprise Avenue, Secaucus, N.J.
   07094
In Canada: Canadian Manda Group, P. O. Box 920, Station U,
   Toronto, Ontario M8Z 5P9
Queries regarding rights and permissions should be addressed to
Carol Publishing Group, 600 Madison Avenue, New York, N.Y. 10022

Carol Publishing Group books are available at special discounts for
bulk purchases, for sales promotions, fund-raising, or educational
purposes. Special editions can be created to specifications.
For details, contact Special Sales Department, Carol Publishing Group,
120 Enterprise Avenue, Secaucus, N.J. 07094

Manufactured in the United States of America

10 9 8 7 6 5 4 3 2 1

Library of Congress Cataloging-in-Publication Data

King, Norman, 1926–
     The almanac of fascinating beginnings : from the academy
   awards to the Xerox machine / by Norman King.
        p.   cm.
     "A Citadel Press book."
     ISBN 0–8065–1549–X
     1. Inventions—History.   I. Title.
   T15.K536   1994
   609—dc20   94–20161
                                                        CIP

To Glenn Gerstell of Milbank Tweed. A brilliant lawyer,
honorable man . . . and very tall.
The kindest, most caring and most decent man I know.
Thank you for being my friend.

Contents

Contents

A

Academy Awards

In 1928, Hollywood was already the film capital of the world, as it remains to this day. Louis B. Mayer, the head of Metro-Goldwyn-Mayer, decided that some self-congratulation was in order. He persuaded other studio heads to join him in founding the Academy of Motion Picture Arts and Sciences and determined to give out awards for excellence—a formula adopted by every country and every entertainment medium since.

The first Awards ceremony was held on May 16, 1929, at the Hollywood Roosevelt Hotel. There were only thirty-six tables, and tension was at a minimum since the winners had known of their good fortune since February. *Wings*, a flying picture that is still something of a special-effects marvel, won for Best Picture. Young Janet Gaynor copped the first Best Actress award for three separate pictures, *Seventh Heaven, Street Angel*, and *Sunrise*. Her acceptance speech, despite the long advance notice, consisted mostly of "Oh my, Oh my." German-born Emil Jannings was named Best Actor for *The Last Command* and *The Way of All Flesh*. But he was on his way back to Germany, fleeing the sudden popularity of talking pictures. He did send a wire, read in his impeccable English accent by swashbuckler Douglas Fairbanks: "I therefore ask you to kindly hand me now already the statuette award to me." This brought down the house.

An American institution, now cherished by hundreds of millions of people around the world for its glamour, its surprises, and its inevitable gaffes, was born that night.

Adhesive Postage Stamp

Early postage stamps had to be affixed using the home glue pot, a messy and time-consuming process. The first adhesive postage stamp, that could simply be licked and slapped on an envelope, was created in America, but it should come as no surprise that it was not introduced by the United States Postal Service. Rather, it was the brainchild of a private mail company in New York City, which began supplying its customers with such stamps in 1842. Despite the unappealing taste of the glue, the idea caught on and we have been enduring sticky tongues ever since.

The U. S. Postal Service tried to use a less sticky glue in the 1980s, but stamps fell off envelopes at an alarming rate, and the old formula was brought back. Peel-and-stick stamps were introduced a few years ago, but because they are not available as commemorative stamps, and cost a few pennies more, they have not caught on in a big way.

Air Brake

In 1872, George Westinghouse perfected and patented the automatic railroad train air brake. Up to that time railroad accidents were considered a normal—if too often mortal—aspect of travel by rail. Brakemen were stationed in each car for the entire length of the train, frantically turning hand-operated wheels like a small army of Charlie Chaplins whenever they heard a warning whistle from the engineer.

Westinghouse was only twenty-three when he invented the air brake, and he went on to devise signal systems to further increase safety. High-speed trains would not have been possible without his contributions. Of course he was just working up a head of steam. He wouldn't get around to championing alternating current and building his utility empire until he was in his mid-thirties.

Air-Cooling

During the summer in the 1920s, customers flocked to department stores and movie houses that had installed the new invention called air-cooling. As time went on, the inventor's name for his electric cooling system metamorphosed into the term air-conditioning, but one of the largest manufacturers of these units still bears the inventor's last name. We can all be grateful to Mr. Willis Carrier.

á la Carte Menu

Restaurants as we know them are a fairly recent development. Inns always served food of some kind for travelers who were staying overnight, but the idea of a place where one might dine out in one's own neighborhood did not develop until well into the eighteenth century, first in Paris and then elsewhere in Europe.

Even then, what one ate was determined by the proprietor. All customers received the same set menu. This was called dining *table d'hôte*, literally meaning "the host's table." The first *á la carte* menus, offering diners a choice of selections for each course, was introduced by the great New York City restaurant Delmonico's, when it opened its opulent doors in the late 1830s. Even Charles Dickens, who, on an earlier trip to America, had had scathing things to say about American food, was impressed by Delmonico's. Other restaurants quickly followed suit in order to compete, and the *á la carte* menu became standard the world over—to the extent that in the 1950s, some restaurants contrived to be "different" by going back to the old *table d'hôte* style.

Alka-Seltzer

The name of the real inventor of what would become Alka-Seltzer is lost, but it is known that he was a young newspaper editor in Elkhart, Indiana. In the late 1920s, during an influenza epidemic, he had his employees drink a combination of aspirin and baking soda dissolved in water. The head of Miles Laboratories happened to visit the newspaper, and was impressed that no one was out sick. The editor explained his homemade formula and Hub Beardsley put his staff to work on devising a marketable product that made use of the combination. The result was Alka-Seltzer, which proved also to be useful in relieving headache and stomach ache, not to mention mitigating the effects of a hangover. It was introduced in 1931, and people have been dropping the tablets in water and listening to them fizz ever since. Of course you can't please everyone. In *Never Give a Sucker an Even Break*, W. C. Fields reacted to a fizzing glass of the stuff by putting his hands over his ears and yowling, "Can't anyone do something about that racket!" But then, Mr. Fields always could make the most out of a hangover.

Aluminum

A metallic element, the most common by weight on earth, aluminum was known to the ancients, but was not scientifically isolated until 1827. Various scientists succeeded in making aluminum, but the process proved very expensive and nobody knew quite what to do with the result. In 1886, however, the American scientist Charles Martin Hall, and a Frenchman, Paul Louis Toussaint Heroult, achieved a virtually simultaneous (and completely independent) breakthrough, producing aluminum by direct electrolysis.

That took care of the process, and brought down the price, but the question of what to use it for remained. Aside from cookware, aluminum was not much used, even though scientists regarded it as a marvel. This is one of those cases in which the invention of something else had to be awaited—in this case the airplane. Suddenly the strong, lightweight, gleaming metal had found its major place in our technological world, serving as the most important element in the construction of the bodies and wings of planes.

America's Cup

In 1851, the greatest yachting race in the history of the world was born with a match race around the Isle of Wight, off the coast of England. It pitted the yacht *America*, owned by John C. Stevens, against the English yacht *Aurora*. The *America* won, and when a new challenge was mounted by the British in 1870, the racing series came to be called the America's Cup. The trophy remained in United States hands through twenty-seven races, but was finally wrested away by Alan Bond's *Australia II* in 1983. The loss of the cup intensified the interest of the public, and subsequent races have been televised in their entirety.

Attempting to win the America's Cup is one of the most expensive hobbies a man can have, and it is no accident that among those who have defended the cup are such famous financial names as J. P. Morgan, Cornelius Vanderbilt, and Henry Sears.

Artificial Heart

The first artificial heart was implanted in a human being (following years of animal research) at the Utah Medical Center in Salt Lake City, on December 1–2, 1982. The chief surgeon was Dr. William C. De Vries and the recipient was Dr. Barney B. Clark of Des Moines, Iowa. The sixty-one-year-old patient survived for 112 days.

The artificial heart used was a Jarvik 7. Developed by Dr. Robert Jarvik, the device, constructed of metal and plastic, is attached to the two upper chambers of the heart as well as the pulmonary artery and the aorta that carry blood from the heart. The heart must be maintained by an external machine to which the patient must remain connected at all times. Because of this limitation, the artificial heart is used primarily to keep a patient alive while a heart transplant donor is found. Even in these situations, there are many risks from side effects, particularly clotting, which can cause strokes.

Assembly Line

In order to keep up with the increasing demand for those newfangled contraptions, horseless carriages, Ransome E. Olds created the assembly line in 1901. The new approach to putting together automobiles enabled him to more than quadruple his factory's output, from 425 cars in 1901 to 2,500 in 1902. Olds became known as "The father of automotive mass production," although many people mistakenly think that it was Henry Ford who invented the assembly line. What Ford did do was to improve upon Olds's idea by installing conveyor belts. That cut the time of manufacturing a Model T from a day and a half to a mere ninety minutes.

Atomic Bomb

The creation of the world's first atomic bomb was a classic example of collaboration between European immigrants and top American scientists. The architects of the controlled nuclear chain reaction that made the bomb possible were Enrico Fermi, the Italian physicist, and Leo Szilard, the Hungarian physicist. Fermi, with his Jewish wife Laura, went directly from receiving the Nobel Prize in Stockholm in 1938 to New York. Szilard fled Nazi Germany in 1933 for England and emigrated to the United States in 1938.

In a squash court beneath the football stadium at the University of Chicago, Fermi, Szilard, and 1923 Nobel Prize winner Arthur H. Compton of the United States, headed the team of scientists working under the code name the Manhattan Project. There, on December 2, 1942, the first controlled chain reaction took place in the graphite pile designed by Fermi. The next step was the construction of a bomb, designed by J. Robert Oppenheimer, at Los Alamos, New Mexico, under the direction of Gen. Leslie R. Groves. The first atomic bomb was exploded near Alamagordo, New Mexico, on July 16, 1945. Just three weeks later, on August 6, a uranium fission bomb called "Fat Boy" was dropped on the Japanese city of Hiroshima, and a new age began.

Automat

In 1885 a new kind of restaurant opened in New York City. Called the Exchange Buffet, it served food to working-class men cafeteria style. Women, however, were not permitted on the premises. The first "all-will-be-served" cafeteria opened in Philadelphia in 1903. It was renamed The Automat, a name picked by Messrs. Horn and Hardart when they opened a new automated cafeteria in Manhattan in 1913. With individual servings behind coin-operated glass doors, the New York Automat was such a success that there were forty in operation throughout New York City by 1939. Automats disappeared at a rapid pace during the 1970s, unable to compete with MacDonald's, Burger King, Wendy's, and the rest. Several generations of children taken to the automat found a thrill that today's fast food chains cannot begin to provide.

B

Bakelite

What did it take to create the first completely synthetic material on earth? Just two fairly simple chemical molecules, phenol and formaldehyde—and many years of experimentation. As early as 1871, chemists started mixing the two chemicals, but did not come up with anything useful, although there was an amber syrup that seemed to hold out possibilities.

The man to come up with answer was Leo Hendrik Baekeland, a Belgian-born chemist who emigrated to the United States in 1889 at the age of thirty-six. In America, he invented a photographic paper called Velox and eventually sold the rights for it to George Eastman for around $800,000 in 1899. He set up a new laboratory and worked on numerous projects, but kept coming back to the phenol/formaldehyde problem over a number of years. It was a difficult test of ingenuity, because so many variables were involved, including amounts, temperature, pressure, and the use of other ingredients. Baekeland finally discovered that to make the amber syrup into something useful, the process must take place in stages, and in June of 1907, he solved the problem of how to use the substance. The resulting synthetic substance made him another fortune, in part because it was a very good insulator and was the perfect material for encasing radio tubes, thus greatly enhansing the marketability of radios.

Band-Aid

The Band-Aid was invented by Earle Dickson in 1920. Adhesive surgical dressings had already been pioneered by Brooklyn-born Robert Johnson, but Dickson reduced the size so that the bandages became practical for home use.

Barbed Wire

Today simply say "barbed wire," and let it go at that. But in the years from 1868 to 1885, no less than 306 patents were issued for different kinds of barbed wire. It was simply wire twisted in different ways, with a barbed end to keep cattle in and strangers out. Yet few inventions of such a simple nature have ever had so profound an effect on the world. Within a decade of the introduction of barbed wire, what we think of as the "Wild West" was gone. The railroads had something to do with that, as did the total subjugation of Native American tribes and the rise of "law-and-order" in the raw frontier towns.

But barbed wire had more to do with changing the nature of the West than did anything else. Great open spaces as far as the eye could see were no longer the hallmark of this vast land. It was parceled out, contained, fenced in with barbed wire. Half a continent was changed in a very few years from untamed wilderness to property. And that property was defined by twisted wire with sharp points.

Baseball

Baseball may still be "the great American pastime," although its heels are being seriously nipped by basketball, but there are a couple of important things it is not: (1) invented by Abner Doubleday and (2) a uniquely American game.

Baseball as we know it, and as it is now played in Central and South America, Japan, and an increasing number of other places around the world, is certainly an American creation, but bat-and-ball games go back to the very beginnings of Anglo-Saxon (British) history. The forerunners of baseball had colorful names like old cats, stool ball, and rounders (which called upon players to round a given number of bases). In fact most of the basic elements of baseball can be found in one or another of these ancient games, but the particular form it took in America is sufficiently different to puzzle the English as much as their national game of cricket puzzles us.

Gen. Abner Doubleday, a certified Civil War hero, got the credit for inventing it—credit that he never claimed for himself—because of a debate about whether baseball had roots in British culture. Doubleday was the perfect symbol for the nationalistic argument that baseball was as American as apple pie, and the myth persists among many people. In fact the first recorded baseball game in the United States was played on June 19, 1846, on the grounds of the romantically named Elysian Fields in not so romantic Hoboken, New Jersey. The original twenty rules used in that game were set down by a man named Alexander Joy Cartwright, who worked for a New York bank and played on one of the two teams that first game day, the

Knickerbocker Base Ball Club. They lost to the New York Nines 23–1.

Basketball

On December 21, 1891, James Naismith, an instructor at the YMCA Training School in Springfield, Massachusetts, threw a soccer ball into play on the gymnasium floor. Only a few minutes earlier peach baskets had been nailed to the balcony at either end of the room. The baskets, bottoms still intact, had been provided by the school janitor. As to rules, Naismith had scribbled out thirteen of them that very morning—twelve of which are still used in the great American game of basketball.

Basketball was an instant success, first with Naismith's students, then at YMCAs across the country. The first public game took place on March 11, 1892, in Springfield. By the following year, the first women's game was played at Smith College in nearby Northampton. A true basketball replaced the soccer ball in 1894. Over the years, the rules changed to make the game more sophisticated, but Naismith's central concept remained intact.

The game was not only popular in America, but spread quickly to countries around the world in a way that football never did and baseball is only beginning to do. It was played for the first time at the Olympics in Hitler's Berlin in 1936. The seventy-five-year-old James Naismith was there to see the American team win the gold medal. The cost of his trip had been raised through a nationwide fund-raising campaign headed by college coaches across the country.

Bifocals

Benjamin Franklin did the world a big favor by inventing bifocals, c. 1760. To be able to read and then walk across a room without bouncing off walls while wearing the same pair of glasses was a big step forward in optics. Bifocals are a prime example of necessity being the mother of invention, since Franklin required such glasses himself.

Biological Pest Control

Biological pest control is all the rage now, an antidote to the overuse of pesticides. There are even ladybug farms, which ship the useful little red bugs out in containers. But biological pest control actually goes back nearly a hundred and fifty years. The first recorded instance of it occurred in 1852, when thousands of sparrows were imported from Germany to help combat a ravenous plant-devouring caterpillar that had been brought into the United States by accident. Our native birds apparently didn't find swallowing this new caterpillar in the least bit tempting, and so birds that regarded it as an especially tasty morsel had to be brought in to do the job.

Birth Control

Birth control had been practiced in various ways around the world from time immemorial, but it was not considered a subject for public discussion when Margaret Higgins Sanger began her crusade on the subject in 1914. A trained nurse, Sanger was appalled by the suffering she witnessed among the poor because of the birth of too many children, and she founded the first birth control clinic in Brooklyn, New York, in 1916. That led to her arrest, as did many of her other efforts. But by the 1920s Sanger had become world famous, and she organized the first international conference on birth control, which took place in 1925.

Black Folk Opera

Many people think of George Gershwin's *Porgy and Bess* as the first folk opera to focus on African American characters. But that isn't true. The first full-length black folk opera was composed in 1911, twenty-four years before Gershwin's masterpiece, by the ragtime composer and pianist Scott Joplin. *Treemonisha* was never given a professional production in Joplin's lifetime, but when his propulsive ragtime songs were revived to broad popularity in the 1970s, attention began to be paid to his other works. Belatedly, but with great success, *Treemonisha* was finally given its due in a splendid production by the Houston Opera in 1976.

Black Hole

In 1964 John Archibald Wheeler added a splendid new term to the astronomical lexicon when he came up with "black hole" to describe a collapsed star. Black holes cannot be observed. Their gravitational field is so great and so concentrated that not even light can escape from them. Thus, they can only be inferred by the intense x-rays they give off.

Although the term is fully accepted by scientists, *black hole* has a ring to it that has led to its being used in many other ways. For example, of the extremely dense coworker who is constantly messing up, it can be said that "he has a mind like a black hole." American English has more such scientific terminology that can be applied to everyday situations than has any other language in history.

Blood Bank

The originator of the blood bank, a medical institution that has been instrumental in the saving of millions of lives, was Dr. Richard Charles Drew, who opened and operated the first one in New York City in 1940.

Why wasn't Dr. Drew allowed to give his own blood to the institution he founded?

You know the answer. He was black.

Boxing Immortality

Cassius Clay, twenty-two, beats Sonny Liston, thirty, in Miami Beach, Florida, on February 25, 1964. Stripped of his title because of his refusal to be inducted into the United States Army, Clay, now called Muhammad Ali, takes the World Heavyweight title again at the age of thirty-two by beating twenty-six-year-old title holder George Forman in Kinshasa, Zaire, on October 30, 1974. Ali reigns for four years but then at the age of thirty-six, is defeated by Leon Spinks, twenty-five, in Las Vegas, on February 15, 1978. But Ali wasn't through yet; he regained title from Spinks in New Orleans on September 15 of the same year, becoming the first and only man to take the title from the reigning champion three separate times.

Brain Surgery

The father of modern brain surgery was Harvey Williams Cushing, born in Cleveland, Ohio, in 1869. After graduating from Yale, he took his M.D. at Harvard Medical School. In 1897 he began a fifteen-year association with Johns Hopkins Hospital in Baltimore, Maryland, as both surgeon and teacher. Brain surgery was not new, but up until that time it was lamentably crude, with a very high mortality rate. With new insights into the anatomy, delicacy, and sometimes amazing resilience of brain tissue, Dr. Cushing pioneered techniques that brought him worldwide acclaim as the foremost neurosurgeon of his time.

In 1912, Cushing became surgeon-in-chief at Peter Bent Brigham hospital in Boston, simultaneously serving as professor of surgery at Harvard Medical School. Until his retirement in 1932, he continued to refine his techniques, wrote many groundbreaking medical monographs and won a Pulitzer Prize for his biography of fellow Johns Hopkins surgeon, Sir William Osler. Most important, he trained scores of young physcians in the techniques of neurosurgery, establishing the discipline as one of the paramount fields of American medicine.

Brassiere

There are numerous sources that list a German inventor named Titslinger as the inventor of the modern bra. Unfortunately, this is one of those stories that is too good to be true, but it is permanently embedded in bra folklore.

The actual creator of the modern bra was a New York socialite named Mary Phelps Jacobs. Various kinds of "breast cloths" had been around since antiquity, of course. These had eventually evolved into full-scale corsets that enveloped the female figure, flattening the stomach, cinching in the waist, and uplifting (or flattening, depending upon the period) the breasts. Mary Jacobs's bra, free of any corset, was sewn together out of handkerchiefs and ribbons. In this case, vanity was the mother of invention. Jacobs had bought a sheer evening dress that her corset showed through, and the "backless bra" was her answer to the problem. It so impressed her friends when she wore it to a party in 1914, that she applied for a patent and began to market her ingenious hand-sewn undergarment. She failed to market it properly, however, and subsequently sold the patent for a mere $1,500.

C

Carbonated Water

In 1811, Joseph Hawkins patented the first machine for turning water fizzy by carbonating it. This bubbly drink took the nation by storm. Brooklyn-born John Mathews, a marketing genius, took the next step by introducing a machine small enough so that it could be used in a store. Known as a "fountain," it gave rise to the institution of the soda fountain. Soon drinks were being flavored with syrups, and then came the fizzy ice cream soda. By some accounts, the number of soda fountains in New York City outstripped the number of bars by the early 1890s!

Carpet Sweeper

Alma and Melville Bissell were allergic to dust. And they were inundated with it when they unpacked boxes shipped to their glassware shop in Grand Rapids, Michigan. In desperation, Melville Bissell experimented with various contraptions that would contain the dust instead of stirring it up the way a broom did. And dust eventually became the mother of a singular invention that the world has been grateful for ever since: the carpet sweeper. The company that bears Melville's last name remains today a giant in the world of dust control. Because the Bissells sneezed so much, we all sneeze a lot less.

Celluloid

Drawing upon earlier experiments in making nitrocellulose (which gave rise to nitrate photographic film), as well as upon some of the work of the English inventor Alexander Parkes, the American John Wesley Hyatt developed a new process and designed machinery for celluloid production. Parker had added oils and solvents to nitrocellulose to form a solid substance through evaporation. Hyatt instead combined nitrocellulose with gum shellac under pressure, and then mixed in camphor. The result was the first modern plastic.

He received a patent in 1869, and in 1871 founded the Celluloid Manufacturing Company. Because celluloid was hard and took color well, it gradually found a number of uses as a substitute for ivory, bone, and shells. Dental plates and billiard balls were among its first applications, to which combs, brushes, and knife handles were soon added.

Chewing Gum

In 1871, Thomas Adams concocted a gum to be chewed. It was made from the dried sap of the Mexican jungle tree, the chiclethe, which was called chicle. Adams was first introduced to it by the Mexican general Antonio Lopez de Santa Anna, the victor at the Alamo. By diluting chicle with hot water and forming it into balls, he produced a more palatable substance. As substitute for chewing tobacco, and perfectly safe for children, it caught on quickly. But there was one complaint: It had no flavor whatsoever. That was taken care of by adding sassafras flavoring in 1875. A national habit was in the making, cemented a decade later by the appearance of an irresistible treat for children known as bubble gum, which had an added emulsifier. Vast efforts on the part of social arbiters ever since to stick chewing gum with the stigma of being lower class have had no effect. It has stuck around in a big way.

Chop Suey

Chop suey may not be as American as apple pie, but there is certainly nothing like it in China. It was invented, created, thrown together—whatever you will—in the early 1860s, maybe by a dishwasher in San Francisco, maybe by a waiter at a Chinese restaurant in Brooklyn. But whoever was the original perpetrator of this dish, it became a huge hit with Americans. A bland mixture of overcooked vegetables (*very* un-Chinese) that always included celery and onion, as well as thin noodles, in a sauce barely touched with soy flavor, it was hardly likely to offend the most timid palate.

Americans have become much more sophisticated about Chinese food in the past twenty-five years, and you won't find chop suey on the menu of any of the upscale restaurants that feature complex and subtle dishes from Chinese provinces that most Americans had never heard of a couple of decades ago. But there's usually someplace in town that still serves it, and you can always buy it in cans if you're desperate. Somebody is eating it, and in fact children often love it.

Cigarette Ads Banned From TV
"Winston tastes good like a cigarette should."
"L. S. M. F. T." (Lucky Strike Means Fine Tobacco).

To anyone over thirty-five, television hardly seems the same without those tobacco company classics. But, in accord with our Puritan background and obsession with health, America quite naturally became the first country to ban cigarette advertising from television in 1971. Harmful to our health—downright dangerous.

Have you ever noticed that the minute cigarette ads got banned, the violence quotient on television began to skyrocket? We traded having "coffin nails" pitched to us in snappy jingles, for more and more bodies in need of coffins. Progress.

Cinerama
Fred Aller of the Paramount Studios special-effects department introduced the first version of what was to become the multi-screen process Cinerama at the 1939 New York World's Fair. At that time, he called it Vitarama. By 1952 he had simplified and improved the process suf-

ficiently to make it feasible to produce the first Cinerama feature, called *This Is Cinerama*. It was still a costly and technically difficult process to work with, however, requiring three cameras filming simultaneously, as well as three projectors and a huge screen with finely calibrated curves. Several features were produced, including the only one with a storyline, *How the West Was Won*, which appeared in 1962. The costs, especially to theaters, prevented Cinerama from ever becoming more than a spectacular fad.

Clipper Ship

The greatest sailing ships ever to grace the seas were the clipper ships of the 1850s. They were largely the work of two brothers, Lauchlan and David McKay. Lauchlan was the theorist, as well as one of the foremost sea captains of his time, while David was the master shipbuilder, possibly the greatest who ever lived. David built the first true clipper, with its three masts and sharp prow, in 1850. It was called the *Stag-Hound*, and every shipowner who saw it wanted something like it.

The most famous of the McKay clippers was the *Flying Cloud*, majestically beautiful and able to make consistent high speeds in any kind of weather. On her maiden voyage she set a new one day record of 374 miles, and she made the trip from New York to San Francisco, around South America, in only eighty-nine days. The *Lightning* was even faster, managing to make a maiden voyage from New York to Liverpool in less than fourteen days. The era of the clipper ships was brief, as steamships took over, but no sailing vessel before them or steamship after could make people's hearts stand still the way the clippers could.

Coca-Cola

It started here and has conquered the world. (When the big international push came in the fifties and sixties, American tourists were stunned to discover that in countries like Spain, a Coke was three times as expensive as a glass of brandy). Coca-Cola (and the fact that all communications with control towers at airports around the world are in English) probably has every bit as much to do with America's special place in the world as all our vaunted military power. It's the little things that count.

The inventor of Coca-Cola was Dr. John S. Pemberton, in 1885. It was sold as a brain tonic, but was poorly received. The formula changed hands three more times before Asa D. Chandler hit upon the idea of carbonating the drink. The bubbles made all the difference, and he struck gold. The Coca-Cola Company was founded in 1892 and has never looked back.

In some ways the company absolutely refuses to look back. There has long been a suspicion that Coca-Cola once contained cocaine. The company resolutely denies it. Corporate and food historians think otherwise. Take your pick.

Color Television

Who actually invented color television? A prototype was built in 1928 by the Scottish inventor John Baird. But the ultimately more important contribution was that of Hungarian-American Peter Carl Goldmark, who developed the first commercially viable color television set in 1940.

Even Goldmark's set had a lot of wrinkles to be rubbed out. Green wrinkles, many of them. Faces had a ghastly greenish tint. Blond hair was mint green. And teeth—well you don't really want to hear about green teeth. Even after the greens were brought under control, there remained a general problem with white, and nobody was allowed to wear it. Nor could the sets handle brilliant colors like orange. As a result, those fortunate enough to own an early color television set were confronted with an absolute sea of pastel pinks and blues. In its infancy color television was in fact redolent of the colors of the nursery. This is to take nothing away from Peter Goldmark. He was one of the true geniuses of his time—the long-playing record was also his baby—and by the mid-1950s the problems had been worked out and the TV color spectrum began to expand in a big way.

Combiner

A consistent theme at the end of the nineteenth century was the development of machines that could do more than one thing at once. A signal example, even to its name, was the combiner, invented in the 1880s by Benjamin Holt. The machine, pulled by large teams of horses, was both a harvester and a thresher, and it was widely used into the early twentieth century. Holt's company developed into the Caterpillar company, virtually synonymous with farming equipment for most of the past century.

Corn Flakes

In welcome contrast to the time-consuming preparation of hot cereals in the morning, Dr. John N. Kellogg came up with the idea of precooked cereals that could be poured straight from box to bowl. Corn flakes were made by partially cooking crushed corn kernels with barley malt, then rolling them flat, baking them in sheets, and finally breaking up the sheets into flakes. Add milk as desired and greet the day with a minimum of fuss. Corn flakes debuted in 1907, offering everyone "The Best to You Each Morning." These days it is "Taste Them Again for the Very First Time," just one more indication of the lasting success of Dr. Kellogg's dry cereal.

Something of a flake himself, Kellogg quickly lost interest in breakfast cereals and turned his attention to yogurt and nut butters. He even penned a paper title "Nuts May Save the Race." Anything is possible.

Cotton Gin

It was a New Englander, Eli Whitney, who changed the landscape of the economically faltering South with the invention of the cotton gin in 1793. The machine separated the fiber from the seed with such rapidity that one man was able to produce fifty pounds of clean cotton a day.

Instead of having a man hold the cotton in one hand while plucking out the fibers from the seeds with the other, a sieve of wires in the new machine held the seeds in place while a revolving drum with hook-shaped wires pulled the cotton fibers from them. A rotating brush then removed the fibers from the hooks.

Whitney did not himself gain much financially from his invention. He received a patent in 1794, but it was given legal protection only from 1807 to 1812, and since he was not able to manufacture machines at anything like the speed necessary to keep up with demand, others stepped in with copies, and it was they who got rich.

Cough Drops

A man whose name is lost to history had invented a formula to ease the pain of sore throats and control coughing. But he was down on his luck, and for five dollars he sold the rights to the formula to two brothers named Smith—James and William. The Smith brothers were candy makers. They added the medicinal formula to a sweet hard candy. A medicine that tasted like candy— Eureka! The public has been sucking on these cough drops (and many imitations) ever since their 1847 debut.

And, yes, those two stern, bearded men you still see on the box are indeed the Smith brothers.

Crackerjack

This ballpark treat beloved of children was not only an American creation, but one that existed long before the first white men came to North America. A special strain of popping corn was hybridized by Native Americans as long ago as A. D. 800. Some New England tribes coated the popcorn with heated maple syrup in order to better preserve it. A crackerjack idea!

Crossword Puzzle

The *New York World* had a Sunday supplement called "FUN" back in 1913, which included a variety of what were called "mental exercises." Asked to come up with a new kind of game, young Arthur Wynne, who had been born in England, drew on a game he remembered from childhood called Magic Square, which simply involved a rearrangement of letters from a given set of words. Wynne played around with this concept and decided to have readers figure out what words were required from a group of numbered clues. He also initiated the concept of "down and across." His puzzle was childishly simple, but readers like it, and as the new fad spread to newspapers across the country, the difficulty of the challenge gradually increased. It was a number of years before some enterprising editor first used the term "crossword puzzle," but the word game is as popular as ever. On the downside, the crossword puzzle created a new form of one-upsmanship: doing it in ink.

D

Democracy

The ideals of democracy are rooted in ancient Greece, even though they were imperfectly realized there, to say the least. Eclipsed for fourteen hundred years, they broke through again in the Magna Carta, signed by Britain's King John in 1215. The writings of Jean Jacques Rousseau in the years immediately preceding the American Revolution focused attention once again on the concept of equality. But if democracy was not a new idea to the American founding fathers, it found its fullest expression yet in Thomas Jefferson's Declaration of Independence.

The political horse-trading that resulted in the original American Constitution left Jefferson uneasy, and he had little to do with hammering it out, but his ideas once again came to the fore in the first ten amendments, the Bill of Rights. Here was the most complete political expression of the rights of man ever to be made into law. For some of the founding fathers, led by Alexander Hamilton, the Bill of Rights went too far. But those rights have been expanded upon ever since, and have been a beacon to the world—from the French Revolution of 1789 to the recent reshaping of the Soviet Union and Eastern Europe.

Deodorant

It was proably inevitable that the deodorant would be invented in America. The peoples of the Mediterranean were never much concerned about body odor, in part because they found it erotic and in part because their bodies were well adapted to summer heat. Northern Europeans seldom experienced very hot weather—unless they traveled to the Mediterranean countries—and the English, in particular, have a history of ignoring body odor that goes back to the court of James the I.

But once they were transferred to America, in which hot weather prevails for months at a time even in the Northeast, the transplanted Europeans suddenly became very conscious of how they smelled. In northern Europe, if you were a member of the aristocracy, you remained one no matter how you smelled, and if you were a member of the lower classes, you remained one even if you were redolent of all the perfumes of Arabia. But America was an upwardly mobile society and small matters were very important to social climbing.

The first deodorant was marketed in the United States in 1888. It was called Mum, and it was a great success with women who believed that "A woman never sweats—she perspires." Mum combined a perfumed cream with zinc. Later products switched to aluminum chloride. It is the metallic ingredients that prevent sweating—and the smell that follows—although exactly how deodorants work is not fully understood to this day. "Underarm products" for women proliferated rapidly, but it wasn't until the 1930s that men jumped—or were pushed—onto the bandwagon.

Dixie Cup

This is a story of a man with ideas about public health, a doll company, and a reluctant public. Hugh Moore spent four years, from 1908 until 1912, perfecting a wax-coated paper cup that could be thrown away after use—the better to stop germs from spreading. People did not like the idea of drinking out of something made of paper. They did not like the wax coating. They were just plain snooty about the whole idea, germs be damned.

Nevertheless, Moore did find a manufacturer, the Dixie Doll Company. They turned our his Health Kup, and people turned up their noses, until Moore got the bright idea of using the company name, calling them Dixie Cups. These days, of course, you would probably have to change the name Dixie Cup to Health Kup to arouse people's interest, but back then it worked. From 1919 on, the Dixie Cup was a great success.

DNA

In 1993, two different men on death row had their convictions overturned and were released from prison because DNA tests, unavailable when they were originally convicted, showed that they could not have committed the crimes for which they had been sentenced. DNA is short for deoxyribonucleic acid, the principal constituent of the chromosomes that determine our heredity. Our DNA is as unique as our fingerprints.

The ability to run DNA tests, not only in criminal cases, but also to determine the paternity of a child or fetus, derives from the work of James Dewey Watson and Francis Crick, who in 1953 first uncovered the "double helix" structure of this complex nucleic acid. In 1968 they received Nobel Prizes for their pioneering work. It took thirty-five years for test results to become sufficiently fool-proof to be accepted as evidence in court, but it seems inevitable that DNA testing will become one of the most important investigative tools of the coming century. And that is only part of the story. The identification of DNA is crucial to the field of molecular biology, including the ability to clone organisms, producing exact replicas. To accomplish that with human beings still resides in the world of science fiction, but it has already been done with primitive organisms.

Drive-in Movie

Richard Hollingshead, Jr., a chemical manufacturer, did something in 1932 that made his neighbors wonder if he'd gone round the bend. He set up a movie screen in front of his garage, rigged up a speaker, and watched a movie from his car. The next year he patented his idea for drive-in movies, and he opened his first theater in Camden, New Jersey. A success, he opened several more, in New Jersey and then in other states. By 1948 there were 820 drive-ins in the United States, and 4,000 by 1958. Television and eventually mall multiplexes gradually reduced the number of drive-ins to a very few, but for two generations "going to the drive-in" was a central experience of growing up in America.

E

Electric Blanket

The electric blanket was not originally conceived as an item for general use. In 1912, the inventor S. D. Russell came up with the idea of a foot-square heating pad, or "blanket," to be used by tuberculosis patients. It cost the then enormous sum of $150. Even when it was introduced to the general public as a full-size blanket intended to offset the chills of winter, it was not readily accepted. The problem, of course, was safety. Early electric blankets had a distressing tendency to short out and cause fires, giving a whole new meaning to the old advice, "Don't smoke in bed."

The solution was to surround the heating elements with nonflammable plastics, but it was not until the 1950s that the electric blanket became a "must have" in American households. For a decade or more electric blankets sold in the millions, but they still sometimes caught fire when improperly used. And there was another built-in drawback. Body temperature fluctuates during the night, and a temperature setting conducive to falling asleep in a chilly room often was too high for the depths of sleep. People found themselves waking in a sweat and having to adjust the temperature. Thus the craze died out, and electric blankets have become less and less common in the average household.

Electric Chair

As the only advanced nation that still has a death penalty, it is extremely fitting that America was the birthplace of the electric chair. Death by electrocution was the brainchild of one Harold P. Brown. Lacking the technical knowledge to bring his idea to fruition, he enlisted the help of Thomas Edison, who provided much of the necessary equipment.

The first man to die in the electric chair was William Kemmler. His execution took place on August 6, 1890, at Auburn State Prison in New York. It took a full eight minutes to finish him off, but nobody protested. He was, after all, a convicted ax murderer.

Electric Razor

Jacob Schick got it into his head that there was a need for an electric razor when he was in the U. S. Army and ran into problems using his standard-issue Gillette when there was no water or no soap. But he faced two major problems in creating a marketable product. The first was technical: there was no motor anything like small enough to be held in the hand. He overcame that problem fairly quickly, and had patented a small-enough motor by 1923.

The second problem was really the larger one, since it was a matter of attitude. Practically everyone said, "Why should I have an electric razor when my bladed one does the job just fine?" Many inventions disappear into the haze of history for just that reason, and many others do not come into their own until a much later date, when a real need for them suddenly opens up—as was the case with aluminum.

But Schick was certain he was onto something, and he finally raised enough money (i.e., dug himself sufficiently deep into debt) to begin manufacturing razors at the then very high price of twenty-five dollars. It was 1931, hardly a good year to try selling something people didn't think they needed when they had trouble paying for the absolute necessities. Schick wouldn't give up. At first, any small profits went toward advertising, which led to bigger profits, which agian went back into advertising. And time was on his side, since the seemingly endless American fascination for gadgets was itself on the rise. By 1937, he was able to sell almost two million electric razors. That he had truly succeeded was made clear by the sudden appearance of competitors.

Elevators Safe for Passengers

Elevators were in operation well before 1852, but they were used only for freight. That was because the cables supporting them had a distressing tendency to snap, causing the whole contraption to go crashing down to the ground floor. An inventor from Vermont, Elisha Graves Otis, solved the problem with a device that prevented the elevator from falling even if the cable did break. The Otis fail-safe mechanism caused metal jaws to lock into place on the guide rails below the elevator car if the speed of descent increased beyond a given rate.

He came up with the device in 1852, but had difficulty persuading builders to use it until 1854, when he demonstrated the safety of his passenger elevator at the Crystal Palace in New York. Three years later the first passenger elevator was in operation at a china store in New York City. His invention had far-reaching consequences: some go so far as to credit him with the concept of the skyscraper, since buildings of more than a few floors were not feasible without passenger elevators.

Etching

The principles of etching were first employed by the Hohokam Tribe of the American West as long ago as A. D. 1000. These Native Americans produced decorative items, presumably worn as jewelry, by first coating shells with pitch. Patterns were then drawn into the pitch, exposing the shell beneath the lines. The shell itself was then put into an acid bath derived from the juices of cacti. Where the surface of the shell was exposed, the acid ate into it, while the pitch protected the remainder of the surface. Once the pattern had been etched into the shell, the pitch was washed off.

The etching process was not discovered in Europe until the middle of the fifteenth century, when smiths who made armor used the same approach to produce designs on metal. Albrecht Dürer further refined the process fifty years later. But the Hohokam had grasped this advanced technique nearly five hundred years earlier.

Evaporated Milk

New Yorker Gail Borden patented a process for the production of evaporated milk in 1856. About half the water content of the milk was removed, not by heating it, as many people assume, but by reducing it in a sealed vacuum. In 1860 Borden opened a factory to manufacture it—a propitious bit of timing, for the outbreak of the Civil War created a real use for the product as a provision for the Union Army. The fact that Union soldiers had evaporated milk, which will keep indefinitely, while the Confederate army was often on the verge of starvation in the later years of the war, was one of many secondary conditions that resulted in a Union victory. After the war, use of evaporated milk spread rapidly, and it soon became a staple in most American households.

Borden went on to develop processes for concentrating fruit juices and other food products, but it was evaporated milk that originally made Borden a household name.

F

Fig Newton

Americans have never been very partial to figs in their fresh state—something about those seeds—although in Italy the fig season is anticipated with sighs usually reserved for romance. James Mitchel, however, found a way to get Americans to consume figs in large quantities. He invented a machine that could make a tight sandwich of a cookie out of rather hard dough. Teaming up with the Kennedy Biscuit Company of Cambridgeport, Massachusetts, he decided that a nutritious fig jam would make a good firm filling. The name of the nearby town of Newton was coopted to market a product that has Americans consuming figs—however much disguised—to this day.

Fish Stick

"Hey, Mom! What's for dinner?"

"Fish sticks."

"Terrific."

Believe it or not, there was a time when the news that it was fish sticks for dinner elicited something other than groans. The American creator of the frozen-food industry, and founder in 1924 of the General Foods Company, Clarence Birdseye, came up with the idea in 1929 of coating even-sized rectangles of fish with batter and breadcrumbs and freezing them for future use. He was basically standardizing an elegant French dish known as *goujonettes de sole* using less expensive fish and a heavier coating. The product was an instant success and became a staple of most American home freezing compartments.

Over time, however, people got tired of them, and because of the health warnings against fried foods in recent years, the market for fish sticks has taken a serious tumble. The summer of 1993 brought the closing of a major Pennsylvania factory producing Mrs. Paul's frozen fish, with production consolidated in another state. One of the reasons: the decline and fall of the fish stick.

Flying Machine

On the morning of December 17, 1903, Wilbur and Orville Wright rolled their hand-built flying machine out onto the sands of Kitty Hawk on the Carolina coast. They alerted the five men stationed at the life-saving station farther down the permanent sandbar. They came over to assist, and thus witnessed one of the most extraordinary moments in recorded history.

At 10:35 A.M., Orville, thirty-two years old, lay face down in the passenger cradle of their machine and slipped the release holding the plane in position. Thirty-five-year-old Wilbur ran alongside as the plane began to pick up speed, steadying one wingtip until the speed of the plane outstripped him. And then it was airborne, and it stayed that way for twelve seconds over a distance of a few hundred feet. The brothers made three more flights that morning, with Wilbur's fourth effort covering 852 feet and lasting just short of a minute. Human beings had just flown an engine-powered heavier-than-air craft for the first time.

While it is true that the Wright Brothers owned a bicycle shop (they had invented a "safety" bicycle) they were far from simple mechanics. Their father was a bishop of the United Brethren Church, and they were encouraged to pursue their scientific experiments from an early age. By the time of Kitty Hawk, they were already well known in scientific circles for their experiments with gliders. Both made great contributions to their work, but Orville was soon recognized as knowing more about aerodynamics than any man alive. It was the application of that knowledge to the controls of the plane that enabled the Wright brothers to win the race against a far better funded rival, Samuel Langley, to be the first men to fly.

Football

Of the three great American sports, the development of football is the least clear. The rules of baseball were codified quite quickly, and basketball, the most original of the three, was clearly defined from the start. American football was an offshoot of the British games of soccer and rugby football. It evolved gradually over the 1870s, chiefly at what are known as the Ivy League colleges of the Northeast, particularly Yale, Harvard, and Princeton, although Rutgers also played an important role.

By the 1890s, American football had become sufficiently popular that many small cities in the East began to organize teams, unconnected with any college, with a loose-knit schedule of games between different towns. The first man to be paid to play football, and thus the first professional player, was William "Pudge" Heffelfinger, who was hired for $500 by the Allegheny Athletic Association to play in a game against the Pittsburgh Athletic Club in November of 1892. Allegheny won. Pudge was essentially a "ringer," but it didn't take long before many teams started hiring pros to give a boost to the local club members. A completely professional football association was first formed in 1920, but it wasn't until 1932 that official league-wide records began to be kept by the National Football League.

Fountain Pen

It is altogether fitting that in his role as inventor Lewis Edson Waterman, a book agent and an insurance salesman, should have created the first practical fountain pen. What better instrument for signing contracts could there be?

He obtained his first patent on the fountain pen in 1884, established the Ideal Pen Company to manufacture it, and went on to incorporate his company as L. E. Waterman in 1887. In 1900 he introduced the self-filling pen.

Essentially, his fountain pen worked on the same principal as a turkey baster with a rubber bulb at the top, except that the compartment was a great deal smaller. By pulling a piston rod set into the side of the pen, the air was pushed out of the rubber compartment, and ink would flow upward to fill the space. The greatest challenge, however, lay in the construction of the nib, which had to be perforated in a particular way to allow the intake of ink, as well as to create air pressure when the pen was in use so that the ink would not flow out again.

The fountain pen was one of those "secondary" inventions that quite probably had a greater effect on the everyday lives of Americans than far more important breakthroughs. The ingenuity of little things we take for granted can be as striking as that required for world-changing inventions like the telephone.

Frankfurter on a Bun

The frankfurter, of course, was a German invention, at first a length of sheep's intestine filled with a mixture of ground meat and spices. But putting it into a bun was an American idea. Who had the idea first? Maybe it was Charles Feltman at his Coney Island restaurant. Or maybe it was Harry Magley Stevens, who was the head of the food staff at the legendary Polo Grounds. Both claimed to be first, and both have had their ardent supporters. Either way, frankfurters in buns were very much a part of New York eating habits by the turn of the century.

If there is an open question about the inventor, there isn't the least doubt about who gets the credit for the term *hot dog.* That honor belongs to T. A. Dorgan, a New York sports cartoonist who drew a dachshund enveloped in bread. This cartoon was anything but amusing to the purveyors of frankfurters, who fretted that the public was taking the term literally, believing that the sausage was made from dog meat. But the popularity of the frankfurter in a bun spread across America, and the term *hot dog* went with it. The packages may say *frankfurter* or *weiner*, but they are hot dogs to just about everyone.

Franklin Stove

According to the judgment of history and of their contemporaries, the two foremost geniuses among the founding fathers were Thomas Jefferson and Benjamin Franklin. But even geniuses make mistakes, and Franklin made a lulu with the stove he invented. It just plain did not work. In one of those examples of being "too clever by half," Franklin designed it so that the smoke came out the bottom. His idea was that the stove would produce more heat, but in fact the fire went out if you looked the other way for ten seconds.

The basic idea was a good one: to build a freestanding cast-iron fireplace that could be situated away from the wall, thus radiating more heat around the room. But Franklin did not fully grasp that heat rises, and that the smoke would have to be removed through a pipe with access to the outside placed *above* the stove. Eventually the stove was redesigned by David R. Rittenhouse and was in wide use by the 1790s. Quite reasonably, he called it a Rittenhouse stove. But legend has its perogatives; the device is known to this day as the Franklin stove.

Freeze-Dried Food

Although it has become an important commercial product line, freeze-dried food was the result of government-sponsored research to devlop lightweight, nutritious, long-lasting, and easily reconstituted meals for use by the military and, with an eye to the future, astronauts. The first successful rations of this kind were created in the 1950s, and there have been many technical improvements since. The process remains complex. It involves rapid freezing, and then removing the water from the food in a vacuum cooled by refrigerant. Once the dehydration has taken place, the remaining water vapor in the chamber is expelled by heating it. These several stages make the process expensive, but the results have been worth the effort, and freeze-dried products are widely used not only by the military and the space program, but also by campers, mountain-climbers, and explorers. Foods are most successfully freeze-dried when cut into small pieces, but more than six hundred different foods can be successfully freeze dried, from chives (which can be found in any supermarket) to meat, fish, and poultry. All it takes to reconstitute the foods is boiling water, and both nutrition and flavor remain intact even if the freeze-dried packet is not used for several years.

Frisbee

A baker named William Russel Frisbee, of Warren, Connecticut, and later of Bridgeport, came up with a clever marketing idea back in the 1870s. He put the family name in relief on the bottom of the light tin pans in which his company's homemade pies were sold. The pans were reusable, but every time a housewife started to bake a pie in one, she would see the name Frisbee and, it was hoped, think, How much easier to buy one.

Eventually Mr. Frisbee's pies were sold throughout much of Connecticut, including New Haven. There, sometime in the 1940s, Yale students began sailing the pie tins through the air and catching them. A decade later, out in California, a flying-saucer enthusiast named Walter Frederick Morrison designed a saucerlike disk for playing catch. It was produced by a company named Wham-O. On a promotional tour of college campuses, the president of Wham-O encountered the pie-plate-tossing craze at Yale. And so the flying saucer from California was renamed after the pie plate from Connecticut.

Frozen Custard

Why would anyone want frozen custard when they could have ice cream?

That's the kind of question to which inventors pay no attention and Thomas Andreas Carvelas of Yonkers, New York, was an inventor. In 1934, he devised a machine that would produce a very creamy cold custard. The machine itself was similar to a hand-cranked ice-cream maker, but with stronger paddles to deal with a thicker mixture containing more eggs and butterfat. Carvelas started selling the confection around the neighborhoods of New York from an old van. People liked it and the business grew.

Why would anyone want a birthday cake made of frozen custard or an Easter frozen custard cake when they could have chocolate layer cake with fudge frosting, or a lemon cake with strawberry icing (in the shape of an Easter bunny if you wanted)?

The man who now called himself Tom Carvel paid no attention to that question either. And his cakes sold a lot more like the proverbial hotcakes than cold ones.

Why would anyone with a lumbering body, stiff arm movements, and a voice somewhere down there in the gravel pits think he could be the best television pitchman for his own product? It'll never work, he was told.

But Tom Carvel made the ads anyway, and his business grew even bigger. And then other people like Frank Perdue and Lee Iacocca followed in his footsteps. It kind of made Tom Carvel look like some sort of genius. Well, he'd known that all along.

Fuller Brush Man

Long before anyone had ever heard the cheerful jingle, "Avon Calling," the prototype of large-scale door-to-door salesmanship had been established by Alfred Carl Fuller. Born in Norway, he emigrated to the United States in 1903, and first began selling brushes door-to-door in 1905. Within five years, he had twenty-five salesman plying various routes. With a true understanding of catering to what customers hadn't yet realized they needed, his business grew into a multimillion-dollar success story that even spawned two movies, one starring Red Skelton, the other Lucille Ball.

Unlike many competitors, Fuller saw to it that his products were top quality. Many of them lasted for years, which meant that customers were far more willing to buy the new products he kept introducing. His was an example of American entrepreneurship at its best.

G

Geography of the Sea

Mathew Fontaine Maury was a man of many talents. Born near Fredricksburg, Virginia, in 1806, Maury became an officer in the United States Navy, and he developed wind and current charts of the Atlantic Ocean that were instrumental in cutting down the sailing time on many routes. He was also director of the U. S. Naval Observatory from 1844 until 1861. But his great accomplishment was the publication of *Physical Geography of the Sea* in 1855.

This work was the first to present the world's oceans as a subject worthy of consideration and investigation on their own special terms. Making use of wind and current charts sent to him from all over the world in response to his earlier work, it was also the first attempt to begin to map the ocean floor—currents, in particular, giving clues as to the ocean depth at various points. Maury was also a consultant on the attempts to lay the first transatlantic cable, which began in 1855 and were finally successful in 1866.

Golf

Although similar games may have been played in very ancient times, golf as we know it began to evolve in

Scotland in the first half of the fifteenth century. Mary, Queen of Scots, was an avid player beginning in girlhood, and the first golf club was founded during her reign, in the year 1552, at St. Andrew's, whose course remains one of the great ones to this day. Mary also gave us the word *caddy*. While being educated in France, she called the youth who chased after her balls "cadet," pronounced "cadday."

The British brought golf to the American colonies in colonial times, but it did not really begin to catch on in the United States until the late 1880s, when John G. Reid of Yonkers, New York, formed the first American club, also named St. Andrew's. The first United States Open was played in 1894, and by the next year there were forty golfing clubs across the country. The Professional Golfers Association was formed in 1916 by the department store tycoon Rodman Wanamaker of Philadelphia. Its popularity has grown ever since.

Grain Elevator

Grain elevators, or silos, stand out against the sky in farming communities across the country. The first one was built in 1842 in Buffalo, New York, where a great many still remain, and soon they were appearing all over the landscape. They were not constructed just to protect grain against the weather, as most people suppose, but for storing grain for considerable periods in order to take advantage of market price fluctuations. The largest silo in the world is located in Wichita, Kansas, boasting a capacity of twenty million bushels.

Architectural historians regard grain elevators as the most completely American of buildings. While a great many still serve their original purpose, architects have turned them into apartment buildings and even hotels, like the Hilton Inn in Akron, Ohio. But they are most stirring when clusters of them appear as far as the eye can see, rising high above the "amber waves of grain."

Grand Slam of Golf

The lawyer born in Atlanta, Georgia, who played golf as an amateur was by 1930 already widely recognized as the greatest golfer in history. Bobby Jones had by that time won the U. S. Amateur Golf Championship four times, the British Amateur twice, the U. S. Open three times, and the British Open twice, as well as dozens of other championships. But in 1930 he astonished the world by winning all four tournaments in the same year. This was the Grand Slam of Golf in those days. (The Masters did not come into being until 1934, and the PGA Championship was played as a match tournament rather than a stroke tournament, and as an amateur, Jones was ineligible anyway.) Jones was not only the first man to win the Grand Slam, he was the only one. Only Ben Hogan, in 1953, managed to win three out of the four of the contemporary Grand Slam events in the same year.

But while the sports world was still buzzing about Jones's incredible achievement, he stunned it again by announcing that he was through with competitive golf forever at the age of twenty-eight. The great amateur was true to his word, although he lived for another forty years.

Grand Slam of Tennis

In 1938, twenty-three-year old John Donald (Don) Budge became the first man to win Wimbledon and the

U.S., French, and Australian Open, the Grand Slam of tennis. The previous year he had won Wimbledon and the U.S. Open, but had not even reached the finals of the French and Australian championships. The only man to win the Grand Slam since was Australian Rod Laver, who did it in both 1962 and 1969.

The first woman to win the Grand Slam was also an American, the great Maureen Conolly, who achieved the feat in 1953. Since then only Margaret Court Smith of Great Britain (1970) and Steffi Graf (1988) of Germany have duplicated that triumph.

Great Escapes

A man went for a swim in the Detroit River in 1906, causing an enormous fuss. First, the month was November, and a hole had to be cut in the ice for him to dive into the water. Second, he was wearing leg chains, body chains, and handcuffs. His name of course was Harry Houdini, probably the greatest magician and escape artist who ever lived.

This particular stunt, which he invented—as he did so many others—almost turned into a catastrophe. The locks were not the problem. Harry had the keys tucked away in his mouth, and he was a master at the contortions necessary to making his escape. But the currents under the ice of the Detroit River swept him away from the hole he had entered by. There was a thin pocket of air between the water and the ice above that he was able to breathe as he tried to find the hole again. After eight tension-filled minutes, he reappeared and was pulled half frozen from the icy waters.

No one created more spectacular stunts and magic tricks than Harry Houdini, and many of those performed by the top magicians of today are in fact simply fresh applications of principles he was the first to lay down. To his great credit, he was also the scourge of fraudulent mediums and spiritualists who bilked people of their money in return for putting them "in touch" with the spirits of their deceased loved ones.

H

Hair Dryer

Q. What do you get when you cross a vacuum cleaner with a blender?

A. A hair dryer.

No, it's not a joke.

The advertisements for early vacuum cleaners suggested that women might dry their hair by using a hose attached to the exhaust end of a vacuum cleaner. It's a perfectly sensible idea, in fact, but the thought of dragging a vacuum cleaner into milady's boudoir didn't have much aesthetic appeal.

But then Stephen J. Poplawski of Racine, Wisconsin, took out a patent on a small motor for a blender. He'd been working on it for many years, inspired in part by the idea of mixing his favorite drink, a milkshake, at home. Two Racine companies, the Racine Universal Motor Company and Hamilton

Beach, got together to produce the first hair dryers. The early dryers blew hot air through a tube, as they do now, but they were three times larger and weighed up to two pounds. They also overheated easily. But they had it all over a vacuum cleaner.

Happy Birthday to You

In 1893, two sisters from Louisville, Kentucky, were given special notice at the Chicago World's Fair in an educational display for their work as principal and teacher at the Louisville Experimental Kindergarten. That same year, they also copyrighted a song they had written to greet their classes at the beginning of each day. The song was called "Good Morning to All," with lyrics by the younger sister, and principal, Patty Smith Hill, and a tune by her elder sister Mildred Hill, who was also a church organist.

Mildred Hill died relatively young, age fifty-seven, in 1916. Patty Smith Hill had gone on to become a distinguished educator and professor at Columbia University Teacher's College. Then in 1924, something happened to their song. It appeared in a songbook put together by one Robert A. Coleman. What's more, he changed the second stanza to begin, "Happy birthday to you."

Within ten years the song was known worldwide as "Happy Birthday to You."

Patty Hill didn't seem to care too much what had happened to her lyrics, but still another Hill sister, Jessica, decided she had had enough when the happy birthday song showed up in the 1934 musical *As Thousands Cheer*, by Irving Berlin. She sued for royalties. She won—her sister

Mildred's melody was deemed the property of the Hill family. That didn't stop the song from being sung around the world, in countless languages, at birthday parties, but at least the Hill sisters got the credit they deserved.

Harvey Wallbanger

No country on earth has contributed so many lethal cocktails with peculiar names to the repertoire of bartenders as the United States. A fairly recent example is the Harvey Wallbanger, concocted in 1970 and so named because it only took one drink to send California surfer Tom Harvey careening off walls.

Harvey Wallbanger
1 oz. vodka
4 oz. orange juice
Pour into collins glass over ice cubes
Stir
Float ½ oz. Galliano over top

Helicopter

The concept of the helicopter goes all the way back to Leonardo da Vinci, who drew a sketch of such a flying machine in 1483. Over the centuries, many attempts were made to build a helicopter, with efforts intensifying after the Wright brothers proved that a heavier-than-air machine could actually fly. The English, French, Spanish, and Germans all had a go at it, but without success. The problem was finally solved by the Russian-born American engineer Igor Sikorsky. After many experiments he found that the combination of a large main rotor, to achieve lift, and a small tail rotor, to control maneuverability, was the answer. Sikorsky had a long history in aeronautics, having designed and flown the first multiengine aircraft in 1913. But it took him until 1939 to work out the difficulties of the helicopter, nearly five hundred years after Leonardo first envisioned his "whirlybird."

Horseless Carriage

In 1879, George B. Selden applied for a patent for a combustion engine fueled by an early form of gasoline. Experiments with combustion engines had been going on since the late 1700s, and Selden himself tried many different fuels from 1875 to 1878. His "liquid hydrocarbon"—he didn't initially call it gasoline—worked. But being a lawyer as well as an inventor, he wanted to avoid having his patent publicly announced until society was ready for a horseless carriage—he knew he was far ahead of his time. So he kept filing amendments, which prevented his patent from being finally acted upon, until 1895. He then sold the patent to an amalgam of companies that were beginning to make automobiles.

Enter Henry Ford. The association that held the patent managed to infuriate Ford with high-handed disdain for his small company, and he decided to challenge the Selden patent. The suit began in 1903 and dragged on for nine years, during which Ford became more and more successful. The patent trial proved messy, with exaggerations on both sides. For instance, it was claimed that Ford had been driving a gas car *before* the original Selden application of 1879. If so, Ford would have had to be a mere sixteen.

In the end, Ford won the suit, with public opinion very much on his side. Selden's patent was never invalidated, however. There is absolutely no question who invented the gas engine and thus the horseless carriage. It was George B. Selden.

Hula Hoop

You may think that the Hula Hoop was a fad born in the 1950s, but in fact people were doing much the same thing with circular hoops made from grape vines and stiff grasses all over the ancient world. It took a couple of Americans, however, to figure out how to make money out of it. Using the word *hula*, which missionaries brought back from the South Pacific around the time of the American Revolution, Richard P. Knerr and Arthur K. Melvin manufactured a plastic hoop in a variety of bright colors. The Hula Hoop was introduced in 1958 and made the two men very rich indeed.

The fad died out in the sixties, but Hula Hoops are now very much with us again thanks to the fitness craze. What better way to grind off those unsightly bulges around the waist and hips.

Hydrogen Bomb

An atom bomb derives its explosive force from nuclear fission, in which uranium is broken apart. A hydrogen bomb creates an even greater explosion by means of nuclear fusion, in which hydrogen gases are slammed together. In order to generate the enormous heat necessary for fusion to occur, an atom bomb must be used as a triggering device to fuse the hydrogen molecules. Because the hydrogen bomb combines the radioactive fallout of the triggering atom bomb with the greater heat of fusion, it is far more devastating.

The first hydrogen bomb was built by the United States and exploded at Enewetok Atoll in the Pacific in November of 1952. The power of the bomb was equivalent to that of ten million tons of TNT. The Soviet Union exploded a hydrogen bomb of its own the following year.

I

Incandescent Lamp

In 1879, after years of research, Thomas Alva Edison produced the first commercially viable incandescent lamp. What came to be called the light bulb changed the face of world, but it's important to note the words *commercially viable*. As early as 1860, the distinguished English inventor Joesph Swann—who had previously made great contributions to the field of photography, including the dry plate—created a functioning incandescent lamp, and later, in 1890, came up with a light bulb that was a considerable improvement on Edison's.

It was Edison who got the greater credit, however, in part because within three years he was overseeing the construction in New York City of the first central electric power plant. Sir Joseph Swann was an inventor of the first order, but Edison was a true genius; his ability to take one invention and follow through on its applications revealed a visionary mind of epochal significance.

Instant Coffee

There have been a lot of men named George Washington down through American history, but aside from the father of our country, the one who probably had the most effect on American life was born in Belgium of American parents. All he did was to invent instant coffee and open the first factory for its production in Brooklyn in 1909.

Brewed coffee was dehydrated in a vacuum, and could be instantly reconstituted with hot water, although it took a great deal of experimenting with the strength of the initial brew to arrive at a balance that retained good flavor.

As with evaporated milk a half century earlier, this product initially met with a good deal of resistance, but just as evaporated milk had found its place as rations for Union soldiers, so instant coffee made life better in the European trenches for U. S. troops during World War I, and went on to vast success in the postwar world.

Interchangeable Parts for Guns

Those who are against gun control should remember that when the founding fathers wrote of the constitutional right to bear arms, there were no guns in the world that were exactly alike. A number of inventors and military men had worked on the idea of guns with interchangeable parts, but it was not until 1798 that Eli Whitney figured out how to actually manufacture such guns.

He made a metal template for each part of the gun, much in the way a dress pattern is laid out on paper. The outlines of the template were then chiseled into the metal to be used for the guns—a time-consuming process that nevertheless paid off. The invention naturally made warfare even bloodier, just as the development of the repeating rifle during the Civil War would do sixty-odd years later.

It is ironic, in this context, that Whitney's invention of the cotton gin—which helped to develop the economic power of the South to the point that civil war could even be contemplated—brought him very little money, since competitors rushed in with improved models. The manufacture of guns with interchangeable parts was the invention that made him his fortune.

Iron Lung

American physicians Philip Drinker and Lewis A. Shaw developed the first practical iron lung in Boston, Massachusetts, in 1928. It was used to treat patients whose breathing muscles and organs were paralyzed, most commonly as a result of polio.

The principle was simple. Except for his head, the patient was fully encased in a metal tank. A rubber collar around the patient's neck prevented air from escaping the tank. An electric pump extracted air from the tank, making the chest and lungs expand because of the falling pressure, which in turn caused air to be drawn into the body through the nose and mouth. As the tank filled with air from the pump once again, the rising pressure would deflate the lungs. The iron lung was a major break-through in terms of prolonging life. Some people were kept alive for many years through its use.

Today, modern life-support machines continue to use a pump, but with advances in technology, the large tank is no longer necessary. Instead a ventilating tube is inserted directly into the patient's windpipe.

J

Jazz

"Jazz." Even on the written page the word jumps out at you. And the syncopated dissonance of jazz has been jumping since the early twentieth century. While some may think of the musical comedy as a purely American art form, it was in fact simply a development of British operetta. Jazz is absolutely American. It, too, has roots, from African American gospel to blues and ragtime, but it developed into something entirely its own. Jazz started in New Orleans, but quickly moved up the Mississippi, gathering steam in Memphis and St. Louis and landing in Chicago in the 1920s.

The early names are legendary: Louis Armstrong, Kid Ory, Jelly Roll Morton, King Morton. Whites, like the cornetist Bix Beiderbecke, took it up, and white bandleaders like Paul Whiteman and Benny Goodman brought a broader audience. The orchestras of Duke Ellington and Count Basie became world famous. And jazz kept right on evolving, through the bop artists Charlie Parker and Dizzy Gillespie to progressive jazz exponents like Stan Getz and Dave Brubeck. Sonny Rollins and John Coltrane ushered in hard bop, and still there was no end to the permutations. Almost all the jazz styles still have their special artists and devoted fans to this day.

And while the cry "rock and roll is here to stay" may be true enough, rock would never have arrived if there hadn't been something called jazz.

K

Kinetophonograph

In the late 1880s, Thomas Edison turned his attention to developing a machine that would "do for the eye what the phonograph did for the ear." The result, in 1889, was the kinetophonograph, which could show film synchronized with a phonograph record. The film projector and the phonograph were started simultaneously by pressing a single button, but if either the film or the record skipped while playing, the results could be unintentionally hilarious. The film used was provided by George Eastman, opening up a new use for his just-invented celluloid film. Edison asked Eastman to provide 35 mm stock, with four precise sprocket holes on the two sides of each frame. That specification remains in force to this day.

At first, Edison's kinetophonograph was designed for use only in a peepshow format, but in the meantime two other inventors had gone on to improve an earlier device of Edison's, developing a projector named for them, the Jenkins-Armat Vitascope. Acquiring rights to the projector, a program of twelve short subjects, many of them lasting only ten seconds, was included on the bill at a New York City vaudeville show in 1896. By 1903, Edison had produced his first "feature," ten minutes long and titled *The Great Train Robbery*. Edison's interest in film gradually waned, in part because of antitrust actions, and he was out of the business by 1917, but he had once again laid some of the important foundations of an entire new entertainment industry.

Kleenex

The invention of Kleenex was not dictated by customer demand or a visionary idea. Instead, it was the result of much corporate head-scratching to find some way to use up a vast surplus of a material that had been designed for other purposes.

World War I had brought on a global shortage of cotton. Kimberly-Clark stepped into the breach with a product called Cellucotton, a cellulose derivative of cotton that was used for wound dressings and the linings of gas masks on the European front. After the war, there was a vast amount of Cellucotton left over, but raw cotton supplies were suddenly more than ample once again. Kimberly-

Clark decided to market a thinner tissue made from Cellucotton and dreamed up the name Kleenex. It was specifically promoted as a coldcream remover, with an advertising campaign that drew on the endorsements of female stars of stage and screen.

The campaign worked, and the new product sold well. But now consumer demand took over. The public, male as well as female, started to use it to blow their noses, throwing the soiled tissue away immediately. Curiously, it took streams of letters to persuade Kimberly-Clark to test market the product as disposable handerkerchief. Thus it was not until 1930, six years after it had been introduced, that Kleenex became identified with its current primary usage, becoming in the process a generic name for tissues no matter what the brand.

L

Laser

Light Amplification by Simulated Emission of Radiation.

Or, in short, Laser.

The concept sounds complicated and in a theoretical sense it is, involving the excitation of atoms to emit photons (a form of energy) that produce an intense beam of light in a reactive process. It is not the energy involved that is crucial, however, but rather the narrowness, and thus the intensity, of the beam of light. All the light waves in a laser beam are of exactly the same frequency and move together absolutely in phase.

The American physicist Theodore H. Maiman invented and demonstrated the first laser in 1960, although it was established in a twenty-year patent suit that some of the underlying patents belonged to another inventor, Gordon Gould. In terms of use, lasers are capable of being adjusted to penetrate steel or to perform the most delicate eye surgery. Aside from their industrial, medical, and research applications, lasers have been used by artists to create holographs, and of course they have become an almost mandatory aspect of rock-show spectacles, with brilliant-colored beams of light cutting through the air, and often changing with the pulse of the music, giving the impression that the mother ship from *Close Encounters of the Third Kind* is about to land.

Laundromat

J. F. Cantrell is hardly a household name—and why should he be, since what he did was to come up with an idea of making it possible for people with small apartments or limited budgets to do something outside their residence that the well off were now able to do at home. Cantrell opened the first laundromat, Washateria, in Forth Worth, Texas, on April 18, 1934. He had bought several electric washing machines, put them in a storefront, and rented them out by the hour.

Liquid Rocket Fuel

The father of the modern rocket is widely believed to be Wernher von Braun, thanks to his great talent for publicity and his highly visible association with NASA. But in fact von Braun was only fourteen when the first liquid-fuel rocket was launched, in Auburn, Massachusetts, by the American physicist Robert Hutchings Goddard in 1926. That breakthrough, one of the most important of the twentieth century, went largely unheralded at the time. Goddard was the first to prove that the gases formed by burning fuel in a combustion container could produce sufficient thrust to lift a rocket when they are expelled at a controlled rate through a nozzle. He went on to design and build high-altitude rockets, and to create the first practical steering system for rockets, along with many other devices. But he was still unsung at the time of his death at seventy-three in 1945. A modest man, he would probably be amazed that his name was eventually given to one of the country's major space centers, the Goddard Space Center near Washington D. C.

Listerine

Nineteenth-century America was virtually awash with patent medicines that were sold without prescription. Exhorbitant claims were made for thousands of potions, elixers, and creams, many of them carried from town to town by traveling salesmen, a practice that gave rise to the derogatory term "snake oil salesman." But one patent medicine caught on in a major way. It was a formula developed by D. Joseph Lawrence of St. Louis, Missouri, and sold as a mouthwash and gargle claimed to "kill germs by millions on contact." Containing a mixture of oils and salts derived from thyme, eucalyptus, and menthol among other ingredients, it is also more than twenty-five percent alcohol.

Lawrence called his product Listerine, capitalizing on the fame of the British surgeon Dr. Joseph Lister, who had been campaigning for santitary operating-room procedures for twenty years. Listerine was to remain the best-selling mouthwash for nearly a century, but received a setback in the 1970s when its advertising claim that it could prevent colds was disproved in extensive tests, bringing on a court order that Warner-Lambert, its present-day manufacturer, spend millions on ads disavowing the claim. Listerine still sells extremely well as a mouthwash, but its image is not what it was.

Long-Playing Record

Peter Carl Goldmark, born in Hungary, trained in Vienna, and a 1933 emigrant to the United States, was one of the foremost figures in the development of modern sound and visual transmission. As soon as he arrived in America, he began working in the laboratories of the Columbia Broadcasting System. There he designed the first practical color television system, which was first used on an experimental basis in 1940. He was the primary genius behind the development of the 33⅓ rpm long-playing record, introduced in 1948, and later went on to develop special cameras for NASA that were used in moon-orbiting satellites and made the high-resolution mapping of the moon possible.

Goldmark was a pioneer by instinct, and he often left the refinement of his inventions to others.

In many ways the long-playing record was simply a refinement of the original wax gramophone record developed by Alexander Graham Bell on the basis of Edison's discoveries about the transference of sound waves to tin foil. But it was a daunting technical challenge because it involved a far more complex ratio between the width of the grooves and the speed of play than was the case with the 78 rpm.

The long-playing record brought untold hours of pleasure to millions of people around the world. With the advent of CDs, the demise of the long-playing record was widely prophesied, but by mid-1993 those prognostications were being hedged as the old familiar vinyl began to show a remarkable upswing in popularity.

M

Machine-Picked Tomato

Today, 90 percent of all tomatoes are picked by machine—which is why most tomatoes are so lacking in flavor and juiciness. Starting in 1947, G. C. Hanna and a team of agriculturists began developing a tomato that would be suitable for machine picking. Once they had succeeded, the machines themselves were developed in the early 1960s. Now you can always lay your hands on a tomato, regardless of the time of year, but salads have never been the same. Everybody complains about the cottony taste and rubbery consistancy of the product. The reason tomatoes marked "vine ripened" are so much more expensive is that they still have to be picked by hand.

Microwave Oven

What do World War II British radar defenses and a candy bar melting in the pocket of an American scientist have to do with each other? The answer lies in the story of the microwave oven. In 1940 British scientists John Randall and Harry Boot invented the cavity magnetron, a device that produced electromagnetic waves that could be tightly focused, so that it could be used like an invisible searchlight to detect Nazi submarines when they surfaced off the coast of England.

The candy bar entered the picture in 1946, when Dr. Percy Spenser was working with a magnetron at the laboratories of the Raytheon Company. A candy bar stashed in his pocket melted into a gooey mess. Following up on this unexpected event, he soon had popcorn bouncing around his laboratory near the tube. Thus it was discovered that a magnetron agitated the molecules of water in food, causing them to first align and then reverse that alignment, generating internal heat.

Raytheon developed its Radar Range using this principle, but it was the size of a small refrigerator and too expensive to be of use except to restaurants. Tappan introduced the first microwave oven designed for home use in

1952, but it, too, cost nearly thirteen hundred dollars. Even at that price people wanted one, and in subsequent years, as the price went down, microwave ovens became a huge commercial success. There were those who resisted—in the mid-1980s Julia Child sniffed that she used hers only for defrosting. With improved browning capabilities, however, microwave ovens began to attract even serious cooks, and these days even *Gourmet* magazine prints microwave recipes.

Mobile Genetic Elements

In 1983, a white-haired American woman in her late seventies was suddenly vaulted from obscurity to worldwide fame. Her name was Barbara McClintock, and she was awarded that year's Nobel prize in Physiology and Medicine for work she had done nearly a half century earlier. As a young researcher in the 1930s, she had spent years cross-breeding varieties of corn. The result of that work showed that genetic elements were not fixed, as had always been supposed, but instead were mobile. In cross-breeding they "migrated" from one area to another, as clearly shown by the different places colored kernels of corn assumed on successive generations of ears of corn.

Barbara McClintock had always worked alone, and had never been a star biologist attached to major university or government research programs. But her work had been published and duly noted by more famous scientists working in the field of genetics. Twenty years before her own Nobel award, in 1962, James D. Watson of Harvard University, and the British biologists Francis H. C. Crick and Maurice H. F. Wilkins had been jointly awarded a

Nobel for "their discoveries concerning the molecular structure of nuclear acids and its significance for information transfer in living material." This work on DNA had brought the three men great fame, and Watson had written a best-selling book, *The Double Helix*. But Barbara McClintock's work underlay theirs, and at last she was accorded the equal honors she deserved.

Monopoly

The game Monopoly was born during the Depression. Unemployed Charles B. Darrow whiled away the hours inventing board games of various kinds. In 1933 he created Monopoly and his Germantown, Pennsylvania, friends, as well as his family, enjoyed playing it so much that they talked him into taking it to the premiere game company of the time, Parker Brothers of Massachusetts. The executives of the company—perhaps because they were so well heeled themselves—rejected it unanimously. They thought it was just plain dull, with rules that were far too complicated. That made Darrow mad, so he took the game to Wanamaker's department store. They thought it was a terrific idea and agreed to carry it in their Philadelphia store. It became an immediate hit, and Parker Brothers belatedly saw the light.

Monopoly's success made Charles Darrow a millionaire as his game became the rage not only in America, but around the world. Versions were developed in nineteen languages, with street names derived from real cities in each country, just as Atlantic City had been the model city for Darrow's original version. Monopoly is still very much with us, and it is the number-one-selling game of the century.

Morse Code

Samuel Finley Breese Morse was in many ways an unlikely man to have have had such an enormous effect on communications. Although he had been educated at Andover and Yale, he had little scientific background. Indeed, his early success came as a portrait painter, the most esteemed young American artist of his time. But even as good a painter as Morse kept running into economic problems: a large project he envisioned for the U. S. Capitol and other public buildings faced political and aesthetic opposition. Thus he was open to other possibilities, and on a voyage back from Europe in 1832 he became fascinated by the new field of electromagnetism.

His early efforts to build a telegraph machine failed. Only because he was willing to turn to others with greater scientific background for help was he able to persevere. Some of these consultants did not get the credit they deserved. Others were were too interested in their own projects to really care about the use he made of their ideas. Finally, Congress appropriated the money to build the first functional telegraph line, between Washington and Baltimore, using electromagnets to receive the dot-and-dash code Morse had developed.

He promised the daughter of a friend the privilege of sending the first message once the line was constructed because she had brought him the news of congressional approval. Her choice of message was brilliant: "What hath God wrought?" Those words tapped out in 1844 in Morse Code were the start of a communications network that would be instrumental in forging the vast reaches of a continent into a cohesive nation.

Mother's Day

Julia Ward Howe, who wrote the words to "The Battle Hymn of the Republic" early in the Civil War, first suggested putting aside a day to honor mothers in 1872. But the idea languished until it was taken up by Anna Jarvis in 1907. Miss Jarvis had had an especially close relationship with her mother and had been heartbroken when she died on the second Sunday in May in 1905. Miss Jarvis suggested to the minister of the Methodist Sunday school in Grafton, West Virginia, where she had gown up, that the anniversary of her mother's death be celebrated as Mother's Day, and the first ceremony took place in Grafton in 1908.

Anna Jarvis then embarked on a letter-writing campaign to make the day a national holiday. The House of Representatives responded with alacrity, but the curmudgeons in the Senate wouldn't go along. One midwestern senator screamed, "Might as well have a Father's Day. Or a Mother-in-Law's Day." Miss Jarvis wrote letters for years to everyone who might aid her cause, and the Senate finally gave in. President Wilson signed the legislation and Mother's Day became official in 1915.

Father's Day had a much tougher time getting established. It was first suggested by Mrs. Sonara Smart Dodd of Spokane, Washington, in 1910, and was celebrated there that year on June 19. Mrs. Dodd's mother had died in childbirth, her father had raised her and her five brothers—it seemed obvious to her that fathers also deserved their own day of celebration. But the Congress, every member of which was male, was leery of such self-congratulatory legislation, and it wasn't until 1972 that President Nixon made Father's Day official.

Motion Picture Review

An early sign that the fledgling motion picture business would be taken seriously was the publication in 1908 of the first movie review. It was by Frank E. Woods and appeared in the *New York Dramatic Mirror*, which covered the stage, vaudeville, and other performances. But it wasn't until ten years later that regular movie reviews became fully established in the nation's newpapers. Movie critics have never had the kind of clout that theater critics have—the ability to close a show or turn one into an unexpected hit. If enough movie critics pan a much-hyped movie they can do it a lot of damage, but often the advance word on a film has already alerted audience to potential hits and misses. Indeed, critics are constantly at odds with the the moviegoing public on the best films and performances of the year. Critics' awards, such as those of the New York Film Critics, do carry a lot of prestige, but they have little effect on the box office. Major Academy Awards, however, can result in millions of dollars of additional box-office receipts.

Mt. Wilson

The man responsible for the building of Mt. Wilson Observatory was George Ellery Hale, born in 1878 to an extremely wealthy Chicago family. Hale had both a brilliant mind and the connections to raise vast sums of money from private donors for scientific and educational projects. Hale had established a reputation as an astronomer by the time he was twenty-three. He had, after all, recently taken the first photographs ever of solar flames through a telescope. So, when he took a trip to Europe with his bride in 1892, he had access to the major figures in the field of astronomy.

In 1903 Hale first climbed six thousand feet to the top Mt. Wilson, near Pasadena, California, and, looking up through the crystal-clear air, determined to build there the greatest observatory the world had ever seen. With the help of men like Andrew Carnegie, who gave ten million dollars, he succeeded. Mt. Wilson became one of the great scientific centers of the world. In the early 1930s, Albert Einstein first came to America specifically to use the data being amassed at Mt. Wilson in furthering his own work.

Although Hale's Mt. Wilson project was of utmost importance in the world of astronomy, he considered his own greatest achievement the part he played in founding the California Institute of Technology, which would go on to become the great rival of his own alma mater, MIT.

N

NFL

Once upon a time there was created in the United
States something called the American Professional Football
Association. Its first champion team, the Akron Pros, was
proclaimed in 1920; the Pros had the best regular season
record, and there were no playoffs. That same year, in
Canton, Ohio, where the NFL Hall of Fame is now located,
the association's name was changed to the National
Football League. It wasn't until twelve years later, in 1932,
that a championship playoff game took place; in it the
Chicago Bears defeated the Spartans of Portsmouth, Ohio.

In 1960 an upstart second league was born, which
called itself the American Football League. The NFL tried
to ignore the AFL at first, but decided to admit its existence and have a playoff in 1966. The contest wasn't called
the Super Bowl until 1969. But with the merger of the two
leagues and the birth of the soon-to-be-called Super Bowl,
the NFL became the National League Conference, the
AFL became the American League Conference, and both
played what continued to be called NFL football.

The NFL, regardless of all these name changes,
remains a uniquely American institution.

Nuclear-Powered Submarine

The first nuclear-powered submarine was the U.S.S. *Nautilis,* named after Captain Nemo's vessel in Jules Verne's *Twenty Thousand Leagues Under the Sea.* The *Nautilus* was launched in 1954. Nuclear submarines are able to remain submerged for almost unlimited periods of time, making them especially difficult to track. In 1960, the U. S. nuclear sub *Triton,* its name taken from the three-pronged staff of the Roman god of the sea, Neptune, completed the first circumnavigation of the world completely below the surface of the sea.

Nylon

An organic chemist named Wallace Carothers, under contract to E. I. Du Pont de Nemours and Company, invented nylon in 1935, after seven years of work. He discovered that when liquid polymers were blown through ultrathin nozzles, they solidified very quickly into fibers thinner than human hair, which could then be stretched into strong threads that could be woven into cloth. But that was just the start of the process that led up to the actual marketing of nylon stockings. It was not until 1938 that Du Pont announced the invention (Carothers had patented it the previous year, shortly before chronic depression led him to commit suicide). The stockings were demonstrated at the new York World's Fair in 1939 and test marketed for another year. By the time they went on sale in stores all over the country on May 15, 1940, Du Pont had invested twenty-seven million dollars in the new product. Women stormed the stores to buy them.

Advertised as "strong as steel and delicate as a spider's web," nylon stockings were initially cared for by women as though they were silk, and did last a long time. But as they became cheaper and produced in greater numbers, women started to treat them as an everyday item, and the "strong as steel" aspect went out the window. They remain ubiquitous to this day. Few woman realize what a miracle they are wearing: each pair is manufactured from a single filament four miles long that is knitted into three million loops.

O

Oil Well

In 1859, most people knew about the existence of oil. It was hard to miss, since pools of it collected in various places on the surface of the ground from Pennsylvania to Texas. The Seneca Indians employed it as a liniment, a use that farmers picked up on. Diluted, it was sold as a patent medicine—Seneca Oil would give rise to the pejorative phrase "snake oil." But mostly it was a nuisance, especially when it came to drilling for salt. There was a lot of salt in western Pennsylvania, and so much oil seeping to the surface that a dank stream near Titusville was called Oil Creek.

There were those who realized that the oil might be useful, however, at least as a lubricant for machinery and possibly as a fuel for lamps. One such man was E. L. Drake, a former conductor on the New Haven Railroad. He leased some land in Titusville and drove interlocking lengths of pipe into the ground. At least Drake would get salt out of it, and maybe he could find the source of the oil. At a depth of seventy-one feet, on August 28, 1859, the pipe went through a fissure in the underlying rock and oil began gushing out. Thus was the world's first oil well drilled in western Pennsylvania (never mention this fact to anyone in Texas).

The well initially produced four hundred gallons a day, which was soon tripled with a better pump. And the liniment of the Seneca Indians would begin to change the world forever within two years, when the first refinery was built at Pittsburgh. The elixir of modern industry and transportation had been found.

On-Demand Video

In 1993 the Hilton Hotel chain began installing a new advance in electronic entertainment, On Demand Video. Thanks to computerized "filing" systems and the enormous number of different signals that can be carried by fiber optic wires, a selection can be transmitted with as much ease as playing a record on a jukebox. A guest may choose from fifty films; with the press of a button, the film appears on screen in only five seconds. It is predicted that within two years On Demand Video will be available throughout the United States, in homes as well as in hotels, becoming as common as color television sets. This is one of many reasons that numerous experts predict the end of the major networks as they exist today.

Organ Transplant

The start of a revolution in medicine occurred on June 17, 1950, at a Chicago hospital called Little Company of Mary. There the American surgeon Richard H. Lawler performed the first human organ transplant. Moments after the donor was declared brain dead, a healthy kidney was removed from her body and washed in a saline solution. The recipient was a forty-nine-year-old woman who shared the donor's blood type and overall physical characteristics, Ruth Tucker. One of her diseased kidneys was removed, and the donor kidney implanted in its place. (The left kidney is usually used in such operations because it has a longer vein and is easier to remove from the donor.) The renal blood vessels and the ureter of the recipient were first clamped and cut, then reattached to the new kidney.

Ruth Tucker lived for five years following her groundbreaking operation, and she died from a coronary thrombosis unconnected with the transplant. In the years since, heart and liver transplants have also become common, but kidney transplants remain the most successful for two reasons. First, one healthy kidney is capable of enlarging to do the work of two kidneys, and second, even if rejection should occur, the patient can be kept alive with dial-

ysis, in which a machine is used on a regular basis to perform substitute kidney function.

P

Pacemaker

In the late 1950s, Dr. Walter Lillehei of the University
of Minnesota made a breakthrough in cardiac treatment
with the use of external pacemakers. An electrode on a
wire through the chest stimulated the heart, with the
power provided by a small battery pack that was worn
around the waist. The advantage of being able to
change the batteries without surgery was offset by the
likelihood of infection at the point where the wire
entered the chest wall. Today external pacemakers are
used only while a patient is waiting to have an internal
pacemaker implanted.

Internal pacemakers are now commonly used to coun-
teract heart problems. The tiny one-ounce devices cause
no discomfort, although care must be taken to avoid mag-
netic detectors used at security checkpoints, since they
disrupt the electronic impulses being sent out by the
pacemaker.

Patterns for Clothes

In the early 1860s, Mrs. Ebenezer Butterick of Sterling,
Massachusetts, was fit to be tied over the vague instruc-

tions that accompanied patterns for dresses in the publications of the period. Her husband leant a sympathetic ear and decided to do somehting about it. He came up with an answer, too: standardized paper patterns.

In 1863, Mr. Butterick began selling his patterns through small millinery shops, and the patterns caught on so quickly that major department stores began carrying the line within months. The gallant Ebenezer gave his wife co-credit for his tailor-made answer to an old problem.

Peanuts

Peanuts have one of the more curious histories among foods commonly eaten around the world. To begin with, they are not a nut, but a legume, like lima beans, soybeans, and lentils. They are native to South America, but they did not get to the United States directly from there. Instead, they were taken to East Africa by Portuguese traders, and then brought to the American South by African slaves.

The peanut was widely shunned by white southerners as weedlike and useless. Then the great black agricultural chemist, George Washington Carver, appeared. Born a slave in 1864, he went on to become the director of agricultural research at Tuskeegee Institute from 1896 to his death in 1943. Carver made invaluable contributions to agricultural science in many areas, including development of the sweet potato, the soybean, and uses for cotton waste. But the myriad uses he found for the peanut were perhaps his greatest contribution. Quite aside from peanut butter, the oil from the legume is used in everything from mayonnaise to shaving cream, from cheese to linoleum. By showing that the peanut and soybean, with their nitrogen-fixing qualities, could restore mineral-poor soil, Carver revolutionized southern agriculture. We would not have had peanut farmer Jimmy Carter for president without the agricultural discoveries of George Washington Carver, and Jimmy Carter was one president who fully recognized Carver's value.

Pentagon

When it was completed in 1943, the Pentagon, the headquarters of the United States Department of Defense in Arlington, Virginia, became the largest office building in the world. It still is. Unlike most modern office buildings, it does not reach for the sky (a skyscraper is much more difficult to make secure). The complex consists of five adjoining buildings occupying a total of thirty-four acres. Because it is so vast, as well as for security reasons, there are very few individuals who have ever seen all of it. Besides, it would be a project worthy of a whole crew of Olympic marathoners to traverse its 6.5 million square feet of enclosed floor space.

Photographic Portrait

In 1839, the New York University physicist and astronomer John W. Draper succeeded in taking the first close-up portrait using the daguerreotype process. The subject was his beautiful daughter, Dorothy Catherine. To people of the time, its most startling feature was that her eyes were open. Early daguerreotypes required such a long exposure that they were primarily used to photograph landscapes. When they were used to photograph people, seated at some distance from the camera, the subjects' eyes appeared closed because they inevitably blinked during the lengthy exposure. Draper, however, had managed to decrease the exposure time, and his daughter had been able to avoid blinking.

Draper also took the first photograph of the moon, in 1840. Whether the eyes of the "man in the moon" were open was impossible to tell, because of the fuzziness of the picture. The blur didn't matter; the photograph represented the beginning of a new era in astronomy.

Plow

Although he never took out a patent, Thomas Jefferson was the first to recognize the potential importance of the iron plow, with its ability to make straight furrows. In 1793, Charles Newbold of New Jersey patented his cast-iron plow. It was not well received, however—farmers were convinced that iron would poison the soil. In subsequent years, dozens of other patents were taken out by various inventors on what they considered new and better plows, but none of these really satisfied farmers either.

Enter John Deere and his steel plow. The trouble with iron plows was that they did not properly throw off the dirt when used in the wet, heavy soil of the Midwest. Deere, a native of Illinois, was a blacksmith by profession. His eye was caught by a discarded steel saw at a sawmill. He began experimenting and produced a steel plow in 1833. It cut straight through the soil, without clogging or stopping. This time farmers were won over. John Deere's steel plow did the job and the price was right, only ten dollars each for his first models. The device was nicknamed "the

singing plow" because of the humming noise it made
while in use, and it became one of the basic items home-
steaders took with them to open up the West.

Polaroid Camera

Edwin Herbert Land had been experimenting with
polarized light since he was a student at Harvard, and in
1932, at the age of twenty-three, he patented his first invention, a plastic material to eliminate glare when mixed with
glass. Several inventions later, in 1947, he produced the
Polaroid Land Camera, a revolutionary device that provided
a finished photograph six minutes after the shutter was
snapped. This required a "package" for each exposure that
contained sixteen layers of dyes and dye developers. The
negative remained, hidden from view by the developed picture, even after the process was complete. Land surpassed
himself with color film in 1962. He fought off the efforts of
other companies to produce copycat versions of his instant
camera, and the Polaroid remains a unique contribution to
the world of photography.

Polio Vaccine

Poliomyeltis, also known as polio or infantile paralysis,
was one of the great scourges of the world until the mid-
1950s. In its acute form, the polio virus could bring on
severe paralysis, and even lead to death when it prevented
the lungs from functioning.

The first vaccine against the disease was developed by
the microbiologist Jonas Edward Salk, while he was a professor at the University of Pittsburgh. His killed-virus vaccine, administered by injection, was successfully tested on

Pittsburgh schoolchildren in 1954. Five years later, Dr. Albert Edward Sabin, who had been born in Russia and migrated to the United States in 1921, perfected a live-virus vaccine that could be taken orally. The work of these two men made possible the virtual eradication of polio from the developed nations of the world.

Polls

The first poll was taken in 1824. Its subject was the presidential election of that year, which involved four major candidates, John Quincy Adams, Andrew Jackson, William H. Crawford, and Henry Clay. The poll was more like the call-in polls on television these days than a true statistical sampling. Ballots were printed in local newspapers in southeastern Pennsylvania, to be returned by mail. The results, published in a Harrisburg newspaper, showed Jackson to have a large lead over the other candidates. However, the poll was taken at the end of July. Jackson's large lead dwindled to a small one, and in the end no candidate could achieve a majority of the electoral vote, throwing the election into the House of Representatives. Jackson had led in the electoral vote, 99 to Adams's 84, with Crawford and Clay bringing up the rear, but the poll of states in the House gave thirteen to Adams, seven to Jackson, and four to Crawford. Thus John Quincy Adams became president, but he was soundly defeated by Jackson four years later.

Polytetrafluoroethylene

This substance sounds like something you'd rather not know about, but companies have a way of copyrighting weird chemical formulas under a name that sticks well enough to become a household word. This particular substance, ironically, is associated with *not sticking,* to the point that its trademark name was applied to President Ronald Reagan because charges against him seemed to simply slide off. That's right, we're talking about Teflon.

Polytetrafluoroethylene was discovered, almost accidentally, in 1938 by a Du Pont chemist named Dr. Roy Plunkett. Teflon, as Du Pont copyrighted it, was put to use in a wide variety of industrial processes where its nonstick quality proved extremely useful. But it was a Frenchman, Mark Gregoire, who conceived of using it on cookware, which was first marketed in Europe under the name T-Fal. In the United States, Du Pont's Teflon was quickly put to the same use—and became a household word in more ways than one.

Potato Chip

The ubiquitous potato, along with corn, was the greatest contribution of the New World to the food supply of people all over the planet. But potatoes were not easily accepted at first. Many people in Europe believed they were poisonous, and even when they did become accepted, they were regarded as food for the poor. That was also the case in America to a surprising degree, and the social standing of the potato wasn't helped by the flood of immigrants from Ireland who came to the United States fleeing the potato famine in their native country.

One day in 1853, in the upper-class spa town of Saratoga, New York, the status of the potato began to rise when a chef named George Crum did something in a fit of annoyance with a fussy customer. He sliced a couple of potatoes very thin and threw them into hot fat. This was intended to make the customer even angrier, but instead these crisp chips caused exclamations of delight, and soon everyone was demanding "Saratoga Chips."

Within months, paper cones filled with the warm chips were being sold on the streets of New York. Slicing the chips thin enough so that they remained crisp even when cooled broadened their popularity still further, and a great American snack was born.

Printing Telegraph

Royal Earl House, despite his grand name, was born in Rockland, Vermont, in 1814. At the age of thirty he demonstrated the first telegraph printing machine, which became widely used. Like Morse code transmitters and receivers, House's machines were controlled by electro-magnets, but they were larger and more cumbersome. A letter typed at the transmitting end caused a carriage to move at the receiving end, taking a position that would allow only the required letter to be typed out. House's machine was eventually merged with a machine invented by David Edward Hughes and patented in 1856. Both men made other important contributions to communications technology. House was the first to use stranded wire for telegraphing, and in 1878, Hughes, while living abroad, invented the microphone.

Prohibition

While there have been communities, and even countries, that have barred the sale of liquor throughout history, in 1919 the United States became the first country to outlaw the free sale of liquor that had prevailed from its earliest days. The Eighteenth Amendment to the Constitution read: "The manufacture, sale, or transportation of intoxicating liquors . . . is hereby prohibited." To put teeth into the amendment, Congress passed the Volstead Act over President Wilson's veto. Volstead put the amount of allowable alcohol so low that even beer and wine could no longer be sold.

Enter the great era of bootlegging, with its violent criminals like Al Capone and its "gentleman" operators like Joseph P. Kennedy. A new vocabulary had to be invented to describe the goings-on over the next thirteen years, giving us such words and phrases as *bathtub gin*, *speakeasy*, and *near beer*. Government attempts to control bootlegging were as fruitless as current drug-enforcement efforts, and the cost of supposedly preventing people from drinking kept growing. In 1933, Congress finally threw in the towel, and the Twenty-first Amendment repealed Prohibition.

Purification of Water

John Wesley Hyatt, the inventor of celluloid, had one of the most wide-ranging minds of any of the great American inventors. Aside from creating the first plastic (celluloid) and later a widely used form of ball bearing, he also developed a water-purification system.

His process, perfected in the 1870s with his brother Isaiah, involved adding a coagulant that collected impurities in the water as it coursed toward the filters he had designed. This allowed for a continuous cleansing process, instead of having to purify the water in a sealed basin for twenty-four hours. What was more, the filters could be cleaned simply by reversing the flow of the current for a short time.

This filtration system was not nearly as important to the eventual development of twentieth-century industry as was plastic, but it was instrumental in efforts to bring a host of diseases under control. In one of those curious intersections of history, modern water-filtration systems use a great deal of plastic—one invention put to use implementing another by the same man.

Q

Quark

Contemporary physics, whether it is dealing with the mysterious outer limits of the known universe, or the subatomic particles that make up all matter, is as peculiar a world as anything Alice ever encountered on the other side of the looking glass. Physicists are acutely aware that their theories concerning kinds of stars that have never been actually observed, like the collapsed stars known as black holes, or invisible building-blocks of matter that haven't yet been captured, have their humorous side. Thus they often turn to literature when naming something that ought to exist, or has to exist, but that remains elusive in any visual or fully measurable sense.

It was in this spirit that the physicist Murray Gell-Mann used *quark* to describe a subatomic particle that he had theorized must exist. The word had been coined by James Joyce for his equally elusive final work *Finnegans Wake*. Physicists have come to believe that the quark not only must exist but is in fact the fundamental unit of matter. The vast (and vastly expensive) supercollider under construction in Texas, until Congress cut off funding for it in late 1993, was to be used in the hunt for the quark, among other experiments.

R

Rabbit Robot

Why in the world would anyone get it into his head to invent a mechanical rabbit? What earthly use could such a device be to anyone? Well, greyhounds liked to chase rabbits, and with the invention of this mechanical lure, modern greyhound racing was born at a modest horse track in New Jersey in 1919. The rabbit-shaped metal lure was attached by a long arm to a steel track on the infield fence. A motorized pulley system whipped the rabbit around the oval at a speed that kept it always about twenty yards ahead of the pursuing dogs. Thanks to Oliver Smith's invention, greyhound racing became a major betting sport in America and Europe.

Radiocarbon Dating

It has only been in the last forty years that archaeologists and biologists have been able to give dates accurate to within two hundred years of the time at which any fossil of a once living organism originally existed. That is because of the work of the American chemist Willard Libby, who had studied cosmic rays for the United States government during World War II. Many scientists were aware that the carbon inherent in all living things on earth undergoes radioactive decay following the death of the organism. But Libby worked out a practical formula for measuring the rate of decay, thus taking a great deal of the guesswork our of fossil dating. Geiger-counter measurements of organic material such as charcoal, wood, and bone are taken to determine the amount of beta-particle emission, which in turn indicates the amount of carbon 14 (14-c) remaining. With Libby's formula these amounts can be used to fix a date when the organic material was still alive. As a result of Libby's work, first published in 1949, sweeping revisions have been made in respect to everything from when the dinosaurs lived to when humans first crossed the Bering Strait from Asia onto the North American continent. The application of Libby's formula has greatly changed our sense of the earth's history and of our own place in it.

Railroad Robbery

In Indiana on May 22, 1868, a year before the completion of the transcontinental railroad, a gang of masked men carried out the world's first robbery of a railroad train. Outlaws had been holding up stagecoaches from time immemorial, and they found it surprisingly easy to transfer their techniques to the iron horse. Holding up trains was far more lucrative than stagecoach robberies, since trains were likely to carry more money and valuables. Americans were fascinated by this new form of outlaw derring-do, and it was no accident that Thomas Edison's first "feature" (it was only ten minutes long) movie, just after the turn of the next century, was called *The Great Train Robbery.*

Reaping Machine

For millennia, the handheld scythe was used by farmers to reap their crops. Thus the horse-drawn reaper that Cyrus Hall McCormack invented in 1831 set the stage for truly epochal changes in agriculture. McCormack's reaper was not just one invention, however, but a group of interlocking inventions that required the presence of one another to function. There were straight, reciprocating knife guards, a reel, a divider, a main wheel drive, and a platform, all working together to to reap wheat and other crops. McCormak's reaper was combined with Jerome I. Case's thresher before the end of the century to create the aptly named combiner, which carried out both tasks simultaneously. Speaking of apt names, Case's middle initial stood for Increase—which is exactly what happened to the crops of American farmers, thanks to his and McCormack's inventions.

Refrigerated Railway Car

Meatpacker George Henry Hammond, fish dealer William Davis, and others saw the possibilities of shipping food in refrigerated cars in the years following the Civil War. A car of dressed beef was shipped by Hammond from the Chicago stockyards to Boston as early as 1867. The beef was discolored, however, because it had been placed directly on the ice, and a prejudice against frozen beef arose in the East that would take some time to overcome.

The first really successful refrigerated car was patented by Joel Tiffany in 1868; he had been experimenting with various approaches to the problem for a decade. By heat-proofing the doors, storing the ice in compartments, and using fans to circulate the cold air, long hauls became possible if the ice was replaced at regular intervals, usually four or five times during a trip from Chicago to New York. The great advantage of refrigerated beef—with Tiffany's improvements it did not have to be actually frozen—was that a single car could carry 50 percent more dressed beef than it could live cattle. The railroads were not happy with this development, but their refusal to allow the use of refrigerated cars backfired when the great meatpacking barons, including Philip Armour, G. F. Swift, Hammond, and others retaliated by forming a monopoly on refrigerated cars forcing the railroads to accept their terms.

Rubber Band

Staplers, glue, tape, and twine all do their jobs, but there is nothing like a rubber band for holding together a bunch of letters or pencils on a temporary basis. Stephen Perry, of a rubber-manufacturing company bearing the family name, came up with this new use for rubber and patented it in 1845. As is the case with many inventions, Perry was simply "fooling around" with rubber compounds when the idea came to him. This is yet another example of the "eureka syndrome" named after the legendary cry of the Greek mathematician Pythagoras, who is supposed to have risen from his bath and rushed naked into the street crying, "Eureka!" (literally "I have found it") when it suddenly came to him how to measure the purity of gold. Rubber bands also have their recreational uses, from propelling model airplanes to snapping across schoolrooms when the teacher's back is turned. Quite clearly, Mr. Perry's simple invention was one for the ages.

S

Safety Pin

In one form or another, the antecedents of the safety pin can be traced back as far as the sixth century B. C. Pins of various designs were used to hold clothes in place, but they were usually as much decorative as functional. The modern safety pin, with its plain design and protected needle end, was the creation of Walter Hunt, who patented it in 1849. The modern pin, of course, is meant not to be seen, unless it is holding together a diaper.

Safety Razor

Until 1903, every man shaved with the same kind of razor that Sweeny Todd, the "Demon Barber of Fleet Street," used to slash the throats of his victims. They were dangerous when sharp, yet keeping them sharp enough to provide a clean shave meant honing them before every use. Fortunately for generations of men, there was a mid-western hardware salesman who didn't like using an old fashioned cut-throat razor at all. His name was King Camp Gillette. He was always on the lookout for products that could be used a few times and then thrown away—which meant that a replacement would be bought, putting money in a salesman's pocket. One day while shaving, it occurred to him that a disposable razor blade would fit the bill perfectly.

He spent six frustrating years trying to devise a holder and blade to fit his vision before he was finally introduced to an engineer named William Nickerson, who had the background to solve the technical problems. A razor blade is made from a steel alloy hardened with chromium, and is only a little thicker than a human hair. Gillette and Nickerson obtained a patent on their new razor in 1901 and it went on sale in 1903. At first there was little interest in the safety razor, but a wealthy friend of Gillette's provided the financial backing necessary for advertising it. By 1908, the product had caught on, and Gillette's company manufactured 300,000 razors and 13 million blades that year.

Sewing Machine

The modern double-thread sewing machine is inevitably associated with Isaac M. Singer. But in fact the real inventor was another Bostonian, a poor machinist named Elias Howe. After several false starts, he patented his machine in September 1846. But he got nowhere trying to interest a manufacturer. The cost of building the machines—in the neighborhood of $300, a small fortune in those days—was regarded as prohibitive. Howe and his family then went to England in 1847, where their economic fortunes only worsened. Upon returning to New York two years later, he found that sewing machines were being sold for only $100, with the important patents held by I. M. Singer. Singer's machine was an improvement on Howe's in a number of respects, but because it produced the same stitch Howe had patented, Howe sued. Singer could have bought Howe off cheaply in an out-of-court settlement, but he was ornery and refused to do so. That cost him in the long run. Howe won his suit and was awarded a variable royalty depending upon technical design features, on every sewing machine that Singer sold. Howe died in 1867, only forty-eight years old, but in the last few years of his life, his royalties amounted to as much as $200,000 a year.

Shopping Mall

As far back as the Middle Ages, most European communities of any size had central food markets, but the concept of a group of stores selling different kinds of merchandise grouped together in a "park" beyond the center of town is a completely American idea. Surprisingly, the first such grouping of stores predates the automobile. The Roland Park Shopping Center opened in Baltimore, Maryland, in 1896. But it wasn't until the late 1920s that suburban shopping centers began to appear, and it took another four decades for the vast shopping malls of the present to become reality. There are clearly millions of people who love shopping malls, although there is a minority who loathe them. As the writer John Malone has put it, "Only Americans could come up with something that gives you agoraphobia and claustrophobia simultaneously." More seriously, social critics and town planners point to shopping malls as one of the greatest single villains in the decay of what used to be called "downtown" in smaller communities across the country.

Showboat

The frontier towns along the Ohio and Mississippi rivers in the 1820s and 1830s were too new and too raw too offer much in the way of cultural amenities. William Chapman, beginning in 1831, filled the gap with his Floating Theater. A riverboat owner who was something of a stagedoor johnny, Chapman saw a way to combine business with pleasure by presenting musical revues as his boat made its usual freight and passenger stops. Within a few years numerous showboats were plying the heartland waters. Though not as grand as those depicted in the movies based on the Edna Ferber's novel and Jerome Kern's musical, the real showboats nevertheless brought a considerable infusion of glamour into the lives of riverfront townspeople.

Silly Putty

The story of American commerce has always been as much one of marketing as it has of invention. A famous example of what can be achieved by taking a product and repackaging it is provided by the stroke of genius that seized a New Haven, Connecticut, store owner named Paul Hodgson. During World War II, the General Electric laboratories had produced a synthetic, pliable rubber that was cheap and useful for a lot of small jobs as a caulking and molding medium. With the war over, a large supply existed, and no one seemed to have any idea what to do with it. Mr. Hodgson came up with one. He bought a large amount, put it into plastic eggs, and sold it to children under the name Silly Putty. Kids could use it to make spiders and snakes, wad on their faces at Halloween, and even lift comic-strip pictures right off the pages of the Sunday funnies. It has been around ever since.

Skyscraper

Thanks to Elisha Otis and his development of a safe passenger elevator, America embarked on a spree of constructing ever taller buildings from 1885 on. The load-bearing iron-frame structure introduced in Chicago that year by William LeBaron Jenney gave new impetus to the American propensity to reach for the skies, and we have never looked back. An engineer, Jenney worked out by trial and error the basic cross-beam structure, with reinforced corners, that would later be adapted for use with steel. Tall, taller, tallest has been the goal, from New York's famous Flatiron Building of 1902, through the Empire State Building, completed in 1931 and for decades the tallest building in the world, to such more recent behemoths as New York's World Trade Center and Chicago's Sears Tower. For now, the economics of office-building construction and environmental concerns have slowed the upward thrust. It seems unlikely that the two-mile-high cities envisioned by such architects as Buckminster Fuller and Paolo Soleri will ever come to pass. But given the American love of the new and the biggest, anything is possible.

Sleeping Car

The first "sleeping car" for railroad trains was introduced on the overnight run from Philadelphia to Baltimore in the 1830s, but a modern park bench would probably offer more comfort. In 1857, Theodore T. Woodruff came up with an "improved" sleeping car that involved three tiers of bunks that folded out of the walls and ceilings. But these were more like coffins than beds, causing no less an authority on travel than Charles Dickens to disparage the accommodations.

It was up to George Mortimer Pullman to change all that. One of ten childern, Pullman was a brilliant entre-preneur who had succeeded in a wide variety of areas, from raising the level of waterfront streets in Chicago (which alleviated flooding) to running a general store in gold-rush Colorado. In his mid-twenties, he had tried to produce a comfortable sleeping car but with little more success than Woodruff.

By the time he was thirty, Pullman realized that he would have to change the dimensions of the car to make it work. That would mean that platforms, bridges, and other railroad dimensions would also have to be adjusted, but he went ahead anyway with his new car, called the *Pioneer*. Just after he had built the Pioneer, the railroads found they had to make many of the roadbed changes his car would require, in order to accommodate the oversize car carrying President Lincoln's body from Washington to Springfield, Illinois.

With his luxurious sleeping cars, and his dining cars— not to mention the covered passageway between them— Pullman came to dominate the railroad car business, and to shape part of the taste of the age.

Social Security

The Social Security Act, passed on August 14, 1935, became one of the most important creations of Franklin Delano Roosevelt's New Deal. It created a federal system of old-age and survivors' insurance to which employee and employer contribute equally. Initially it was 1 percent of salary, and from the funds collected, the government would pay those retiring at sixty-five a monthly pension that ranged up to seventy-five dollars. The act also provided grants to states to help pay for pensions for individuals who did not come under the federal plan. In addition, a linked federal and state system of unemployment compensation was created, and special provisions were made to help dependent mothers, the crippled, the blind, and others in need.

The entitlements provided by Social Security and subsequent programs such as Medicare and Medicaid have become tightly woven into the fabric of American society, but because they also contribute to the federal deficit, they remain a hot-button political issue, as neither Roosevelt nor the original congressional authors of the act could likely have foreseen.

Soda Straw

Harry M. Stevens, the young man from England with very American ideas, not only was instrumental in popularizing the hot dog at the Polo Grounds for New York Giants baseball games, but also came up with another innovative idea: the soda straw. Concessionaire *extraordinaire*, Harry noticed that baseball fans had to take their eyes off the field when they raised a soda-pop bottle to quench their thirst. Given how suddenly things could happen in a baseball game, the fans could be missing out on crucial moments. Harry was aware that reed tubes for sucking up liquids had been in use since the days of the ancient Egyptians, and he enlisted a paper manufacturer to produce straws by forming paper around metal dowels. He started inserting one in every bottle of pop he sold. With a straw to suck on, the eye could be kept on the ball at all times.

Straws soon moved from the Polo Grounds to the corner store. Mothers liked them because they seemed more sanitary than a glass that had been drunk from by others and simply rinsed out before being refilled. At the turn of the century young people found them fun, and courting couples were known to suck on two straws in the same drink while "spooning."

Solo Transatlantic Flight

On May 20, 1927, Charles Augustus Lindbergh took off from Roosevelt Field on Long Island in his *Spirit of St. Louis* with the intent of flying solo across the Atlantic to Europe.

Backed by a group of St. Louis businessmen, Lindbergh had bought a 5,250-pound monoplane equipped with a new air-cooled Wright Whirlwind engine. Because of a $25,000 prize that had originally been offered in 1919 by a New York hotel owner named Raymond Orteig, several previous attempts had been made without any success. Improved aeronautic technology had enabled a number of new efforts to be mounted, and in the month before Lindbergh's flight, four aviators had been killed trying, two had been injured, and another two had disappeared over the Atlantic.

Very few believed Lindbergh had even a remote chance of succeeding. But as he approached the French coast, word spread of a potentially historic moment, and thousands of people rushed to Le Bourget airport in anticipation of at least seeing him crash. When he landed safely, the crowds erupted in pandemonium. Lindbergh was received in Paris like a conquering hero, but that was only a taste of the adulation that greeted him on his return to the United States. His flight of 3,610 miles had lasted thirty-three hours and twenty-nine and one-half minutes, and made him for some time to come the most famous human being on the planet.

Sound Barrier

The X-1 was a mere thirty-one feet long with a wingspan of only twenty-eight feet. It didn't look like much, but it was something very special, a rocket plane. Named by its pilot, Captain Charles "Chuck" Yeager, for his wife, the *Glamorous Glennis* was scheduled to make an attempt on October 14, 1947, to break the sound barrier. The speed of sound at sea level is 760 mph, but it decreases somewhat at higher altitudes, and Yeager would only have to fly at around 700 mph to achieve supersonic speed.

We are now entirely used to hearing the loud booms in the sky when jets break the sound barrier, and to the wonder of watching exhibition jets like the Blue Angels vault across the heavens well ahead of the wake of sound they leave behind. But what Chuck Yeager was trying to do was revolutionary, and very dangerous. There was always a chance that when he started the rockets on his plane at about forty thousand feet, the fuel tanks would explode.

The development of jets that could take off from the ground and break the sound barrier was years in the future, and *Glamorous Glennis* would be taken aloft by a B-29 Superfortress and dropped from its bomb bay. To add to Yeager's difficulties, he had broken two ribs in a riding accident two days earlier and the use of his right arm was much impeded—a fact he managed to hide from his superiors.

In the end, everything went as planned. Chuck Yeager, at the age of twenty-four, became the first man to fly at supersonic speed. And by doing so, he took one of the first steps toward our eventual exploration of space.

Sound Picture

Iowa-born Lee De Forest introduced the idea of talking motion pictures in 1923. Thomas Edison had earlier combined film with a phonograph recording, but this could lead to unintended hilarity when the phonograph skipped ahead of the screen images. De Forest was the first to develop a process that recorded the soundtrack directly onto the film itself. The sound was initially converted to electricity through a microphone, with the variations in current indicated by the changing intensity of a lamp. The film had two tracks, one for sound and the other for the picture. The sound strip recorded the variations in the intensity of the lamp, producing a pattern of squiggles that were electronically converted back to sound by the projector. But the Hollywood studios were not initially taken with the idea. The public seemed satisfied with silent movies; moreover, the cost of sound movies would be much greater. Studio magnates were all too aware that some of the biggest movie stars had voices that would leave audiences in shock.

When sound did prove popular—with Al Jolson's *The Jazz Singer* as the breakthrough—the studios rushed headlong into the sound era. But due to the power of Western Electric and other corporate giants, Lee De Forest found himself outmaneuvered. He never did receive the recognition or the financial rewards he deserved.

Space Shuttle

The United States took a major leap ahead of the rival Soviet space program with the successful launch of *Columbia* on April 12, 1981. With a length of 122.2 feet and a wingspan of 78.6 feet, the *Columbia* and its sister ships are capable of carrying as many as eight astronauts (which *Challenger* did in 1986) and large payloads such as the Hubble telescope. But what made the space shuttle a particularly notable achievement was that it could return to earth like a conventional plane and land on a runway. The shuttle was the first reusable space vehicle, and has remained unchallenged in that regard.

Stealth Bomber

Few American military developments since the atomic bomb were as cloaked in secrecy as was the Stealth bomber. The bomber was conceived as a plane that could be "cloaked" to make it undetectable to radar. As the years dragged on, however, serious questions came up. Was the plane really capable of hiding from radar? Was it truly fly- able?

There were crashes during testing. And perhaps most of all, critics questioned whether the secrecy surrounding the plane served primarily as a matter of national security, or as a way of keeping the public from finding out how much these strange-shaped black planes actually cost. In the end it became known that each plane cost in the neighborhood of $800 million, and Congress scaled back the number to be built.

The Stealth was finally flown in combat during the Gulf War of 1990. None was shot down, but questions remain about its effectiveness as a bomber. All in all, there are still many experts who feel that the building of the Stealth is an example of American ingenuity run amok.

Steamboat

No, the inventor of the steamboat was not Robert Fulton. It was John Fitch, who launched his first boat on the Delaware River near Philadelphia in 1787. He never got the recognition he deserved, and died a pauper in 1798.

Fulton is so often mistakenly given full credit because he was the first to demonstrate the commercial value of the steamboat when in 1807 his *Clermont* made the 150-mile passage from New York City to Albany in thirty-two hours. It was also true that he had made enormous improvements that would win the day even though most people thought steamboats utterly impractical, even laughable. In fact, Fulton himself died embittered and frustrated.

Despite the successful round-trip voyage of the *Clermont*, the public expressed little interest in it. Once that interest dramatically increased, competitors ruthlessly infringed on Fulton's patents and he was forced to spend as much time in the courts as building ships. At the time of his death in 1815, at the age of fifty, Fulton had no idea that his story would become an American legend and that he would be credited with far more than he ever would have claimed for himself.

Stereophonic Sound

The recording industry in the United States has always had a special genius for introducing an improvement in sound that also requires entirely new equipment to reproduce the enhanced fidelity, which means additional profits on two different fronts. This double parlay was achieved with the introduction of the long-playing record, with the advent of stereophonic sound, and most recently with CDs.

While there are a few critics who think the sound of CDs is *too* perfect, resulting in a kind of sterility, there was no such objection in 1961, when stereophonic sound was first marketed. The new system required recording two separate tracks of sound, one on the left and the other on the right. Separate loudspeakers picked up the different electronic signals, creating the illusion of the "live" sound experienced in a concert hall. This was obviously an enormous improvement over monaural sound. Pioneered by Zenith in cooperation with General Electric, stereo proved an immediate success technically and as a marketing venture.

Stewardess

A registered nurse named Ellen Church became the first airline stewardess in 1930, aboard a twelve-passenger Boeing 80. It may annoy some feminists, but the term *stewardess* was not the creation of a man. Ms. Church chose it herself, feminizing the word used for room attendants on ships. Her main job was to keep passengers calm—in those days the majority of people were flying for the first time. She also passed out chewing gum to help alleviate the pressure in the ears that bothered many people during takeoff. It wasn't until 1936 that flight attendants, to use our contemporary designation, started serving meals aboard planes. Which American airline was the first to try to make us gag at thirty thousand feet is a matter of dispute.

Submachine Gun

The rapid-firing but bulky machine gun was widely used in World War I and has continued to be an important weapon. But in 1920, a retired army officer named John T. Thompson patented a small, lightweight, portable rapid-firing weapon, the submachine gun. It was an automatic weapon—while pressure was maintained on the trigger, the gun would keep firing so long as bullets remained in the magazine or ammunition-feeding device. This weapon was also called the "tommygun," after its inventor, and became a favorite of gangsters like Al Capone during Prohibition. New generations of the submachine gun continue to cause havoc on the streets of America to this day.

Submarine

In the same year that Saybrooke, Connecticut, native David Bushnell graduated from Yale, 1775, he began construction of a submarine. It was a small submersible that could carry only one man, who had to propel the craft manually. During the first year of the Revolutionary War, Bushnell made several attempts to destroy British ships with "Bushnell's Turtle," as it was nicknamed, but without success. Nevertheless, this pioneering effort did garner him the honor of being regarded as "the father of the submarine."

At the order of the French in 1801, Robert Fulton designed and built the second submarine ever constructed. It was tested in the harbor at Brest and successfully blew up an unmanned Danish brig. Fulton's *Nautilus* was a far more sophisticated vessel than Bushnell's. It could carry four men, submerge to a depth of twenty-five feet, and remain under water for four and a half hours. Virtually all the elements of modern submarines, from conning tower to ballast tanks, were features of the *Nautilus.*

In operation against the British navy, however, the *Nautilus* failed to blow up any ships. But the British, recalling the threat posed by Bushnell's submarine, took great pains to stay far away from Fulton's improved model, and to Fulton that was in itself a success. The French, however, were unimpressed, and Fulton went back to America and turned his attention to steamboats.

Superfortress B-29

These brilliantly designed planes were the greatest triumph of the American war manufacturing effort during World War II. Although both the British and the Germans produced some extraordinary aircraft in the course of the war, none was ultimately more important than the American B-29s. With their three-thousand-mile range and four-ton bomb capacity, they were without question the most important element in achieving victory in the

Pacific. From their introduction in late 1943, they gave the Allies a reach between the scattered islands of the Pacific that made possible the retaking of one Japanese-held base after another. And, finally, of course, it was the *Enola Gay*—a stripped-down B-29 named by its pilot after his mother—that dropped the first atomic bomb on Hiroshima on August 6, 1945.

More than sixteen million Americans, in factories across the country, played some part in the manufacture of the B-29s. Never before in the history of the world had there been such a concerted manufacturing effort.

Suspension Bridge

The world's first suspension bridge was built by James Finley in Lancaster County, Pennsylvania, in 1800. It was only a forty-foot span, but it was suspended by iron chains from towers built on either bank of the river.

The prototype of the modern suspension bridge was the Brooklyn Bridge, begun by John A. Roebling and completed by his son, W. A. Roebling, in 1883. John Roebling had constructed earlier wire-cable bridges, but the length of the Brooklyn span and its height above the water necessitated a new approach. He sank pylons deep into the bed of the East River and constructed towers around them in a way that had never before been attempted. It was very dangerous work, and W. A. Roebling in fact suffered the bends while directing operations under the river. During the last year of construction he watched from an apartment on Brooklyn Heights while his wife carried his instructions to the work crews.

The techniques he and his father pioneered have been used in bridge building ever since.

Swimming the Channel

On August 6, 1926, Gertrude Ederle, the eighteen-year-old daughter of a small New York City delicatessen owner, became the first woman to successfully swim the English Channel.

In the previous fifty-one years, since the English swimmer Matthew Webb had first managed the feat, only five other men had done it, although hundreds had tried. Trudy, as she was called, had decided to swim from Cape Fris-Nez, France, to the small port of Deal, England. The Channel waters were so rough on the day she had chosen that steamer crossings were canceled. Trudy paid no attention. Not only did she complete the crossing, she did so in fourteen hours and thirty-one minutes, thus not only becoming the first woman to succeed at the great challenge, but doing it faster than any man before her.

Trudy Ederle's hearing was permanently affected by the pounding she took in the Channel that day, but not to the extent that she couldn't hear the cheers of the nearly one million people who thronged the streets of New York to welcome her home.

T

Technicolor

Technicolor film goes back a lot further than most people realize. The process was invented during World War I by Herbert T. Kalmus and Daniel F. Comstock. It was initially only a two-color system, utilizing red and green. First employed in a film produced by the inventors, *The Gulf Between* in 1917, improvements were made over the years that led to the feature film, *The Black Pirate*, starring Douglas Fairbanks, in 1925. There were other silent films made in color, but it was an expensive process and still impractical, and audiences lost interest.

In 1932, however, Kalmus, whose Technicolor company had been incorporated in 1922, made an enormous step forward by introducing a three-color process. This process was first used in *Becky Sharp* in 1935, and by the golden year of 1939, several now-classic movies were released in color, including *Gone with the Wind*, *The Wizard of Oz*, *The Private Lives of Elizabeth and Essex*, and *Drums Along the Mohawk*. Technicolor became increasingly popular, especially for costume dramas and musicals, but black-and-white held its own for a long time, remaining the preferred medium for serious drama. It was not until 1956 that all five Best Picture nominees, *Around the World in 80 Days, Friendly Persuasion, Giant, The King and I,* and *The Ten Commandments* were in Technicolor. While other color processes have made their mark, Technicolor remains by far the most used.

Telephone

Like his father and grandfather before him, Alexander Graham Bell was an elocutionist, teaching diction and voice projection. His father's eminence in the field was such that George Bernard Shaw borrowed his ideas and put them in the mouth of Professor Henry Higgins in *Pygmalion.* You may say, "Well, that's interesting, but what does it have to do with inventing the telephone?" In fact it has a great deal of bearing on Bell's triumph.

Born in Edinburgh in 1847, Bell came to the United States in 1871 to teach in the field of vocal physiology, a position that had originally been offered to his father at the Boston School for the Deaf. As a result he met a number of wealthy Bostonians, who found him engaging and became interested in his ambitions as an inventor. His chief aim was to invent a method of transmitting several messages simultaneously by telegraph. Almost accidentally, while working with an experimental device, he stumbled onto the essence of the telephone. While telegraphy required pulses of current, a telephone would require a continuous one to prevent the sound of the voice from continually breaking up. Because of his training in elocution and his acoustic work dealing with his deaf students, Bell grasped this difference, one that could easily have made no impression on someone who was merely an electrician.

Bell had his assistant Thomas A. Watson construct a device that did indeed transmit the human voice from one place to another. He first demonstrated it publicly a few months later at the Centennial Exhibition in Philadelphia, reciting Hamlet's "To be or not to be" speech from another room. His patent for the telephone became the most valuable ever issued in the United Sates.

Television Remote Control

The TV remote control was invented as far back as 1956 by Dr. Robert Adler. The press of a button sent an electronic signal to a transistor inside the television set that was calibrated to change the channel. But the early models had to be attached to the set with wires stretched across the viewing space, and even when wireless versions came on the market, they remained expensive. It wasn't until 1982 that the device became a standard accessory for television sets. Dr. Adler does not mind being accused of creating the "couch potato" as a byproduct of his device. But there are those—spouses of people who are inveterate "zappers," switching channels incessantly—who wish that the remote control had never been created. There are couples who, to keep their marriages afloat, never watch television in the same room unless they have rented a movie and it has been formally agreed that it will be watched all the way through.

Tennis

Pictorial evidence shows that a form of tennis, played with the palms of the hands rather than racquets, existed as far back as twelfth-century France. It was so widely played in England by the Elizabethan era that William Shakespeare made reference to the game in several plays, including *Henry V*, and John Webster wrote in *The Duchess of Malfi* that "We are merely the stars' tennis balls, struck and bandied/ Which way it please them."

Modern tennis, initially called lawn tennis, was introduced in England in 1873 by a Major Walter C. Wingfield. The first matches at Wimbledon were held in 1877. By that time the game had already been introudced to America by Maria Ewing Outerbridge of Staten Island, New York, still in her teens, who had seen the game played while on a Bermuda vacation in 1874. She interested her brothers and friends in tennis and the game quickly became popular. The United States Lawn Tennis Association was formed in 1881. National championship matches were held for the first time that same year in Newport, Rhode Island. In 1915 the event was moved to Forest Hills, New York. The game was given a major international push in 1900, when a wealthy American named Dwight F. Davis established the Davis Cup, awarded to the surviving national team after a series of elimination tournaments.

Toll House Cookie

There really was a Toll House. It was originally built in 1709 on the outskirts of Whitman, Massachusetts, and served as a way station where tolls were collected, horses were changed, and stagecoach passengers could get a meal. Eventually the building was acquired by a New England woman named Ruth Wakefield, who turned it into an inn. An inventive cook, Mrs. Wakefield decided to chop up chocolate bars and add the bits to her butter cookies. The chewy delights, which she called Toll House Inn cookies, were an immediate hit, and word of their existence spread very quickly, soon reaching the Nestlé Company. In exchange for a lifetime supply of chocolate, Mrs. Wakefield allowed the company to print her recipe on their chocolate-bar wrapper.

Curiously, it wasn't until the end of the decade, in 1939, that Nestlé fully capitalized on the recipe by creating Toll House chocolate morsels. The enduring popularity of Mrs. Wakefield's culinary invention was evident in the 1992 presidential campaign, when both Barbara Bush and Hillary Rodham Clinton were roped into providing the hungry electorate with their special variations on the original recipe.

Toothbrush Revolution

If you were born after World War II, this question has probably never occurred to you: before nylon was invented, what material was used to make the bristles in toothbrushes? The answer is hogs' hair. But starting in 1938, everyone, including the hogs, got a break. Quite aside from any aesthetic factors, hogs'-hair bristles tended to both fall out of the toothbrush *and* get stuck between the brusher's teeth.

You might think that dentists immediately went hog-wild in their enthusiasm for the new artificial bristles, but in fact the early ones were so stiff that they often caused gums to bleed. It was another decade before a softer nylon bristle was perfected, but from then on dental hygiene showed a steady improvement in the United

States. Contrary to popular assumptions, however, nylon bristles do not last forever. Most dentists recommend getting a new brush at least once a year.

Tootsie Roll

Would you give anything, now and again, to lay your hands on a package of Clara Rolls? No? Fortunately, Clara, the young daughter of a New York candymaker named Leo Hirschfield, also had a nickname. And when the proud papa came up with a new confection in 1896, he wanted to name it for his daughter, but had the good marketing sense to use her nickname, Tootsie. The taffylike mixture of chocolate, sugar, and corn syrup can still be found everywhere.

Transcontinental Railroad

It is difficult for us to imagine what excitement and awe the completion of the Transcontinental Railroad aroused in Americans of the late 1860s. Simply the discovery of a practical route for the railway to follow was regarded as extraordinary, given the difficulties encountered by earlier explorers in crossing the Sierras. The Union Pacific line was built westward out of Omaha, Nebraska, a site that had been chosen by President Lincoln. The Central Pacific line was the more difficult to build, moving eastward from California over the Sierras. Chinese workers in vast numbers did the dangerous blasting work, using nitroglycerin, to create the tunnels through the mountains.

A reporter who was on the scene in Nevada, on the Union Pacific end, wired this dispatch: "Five men to the

500-pound rail, 28 to 30 spikes to the rail, three blows to the spike, two pairs of rails to the minute, 400 rails to the mile—and half a continent to go . . ."

At last the great task was done. On May 8, 1869, the two lines were linked up at Promontory Point in Utah. For the first time in history, an entire continent could be traversed by rail, from one sea to the other. One commentator summed it up by noting that Americans "love to annihilate the magnificent distances."

Tupperware

Earl Tupper knew immediately that polyethelyne, the new plastic that was formulated in 1942, was exactly what he had been looking for over the past several years. A Du Pont chemist, Tupper saw that the pliable, attractive, and very long-lasting synthetic polymer was the right material for a host of home products. He began by producing a bathroom drinking glass available in a rainbow of colors and quickly moved on to his famous lidded bowls.

His products were originally sold in retail stores, but he had another marketing idea that would make him a multimillionaire. His Tupperware parties proved to be an enormous success, and fit in perfectly with the new mobility of Americans in the postwar era. Wherever Americans moved, and they moved more with every passing year, they would find a Tupperware party where housewives could meet new neighbors—and of course purchase some more of Earl Tupper's extremely useful products. Comedians loved to joke about Tupperware parties, but that just provided free publicity. By 1958, Mr. Tupper was able to sell his company for approximately nine million dollars and retire for life.

TV Guide

The magazine that debuted in April 1953 was small, but the editors had made a very clever choice for its first cover: Desiderio Alberto Arnaz IV, the three-month-old son of Lucille Ball and Desi Arnaz, stars of television's most popular show ever, *I Love Lucy.* The brainchild of Walter H. Annenberg, who would go on to become one of the richest men in America and ambassador to Great Britain under Ronald Reagan, *TV Guide* was a success from the start. Initially produced in ten different regional editions, it now appears in more than one hundred each week. Often surprisingly outspoken for a mass-market entertainment magazine, *TV Guide* has not merely kept the public informed about what is on the air, but, true to its name, guided viewers' choices in an attempt to bolster exceptional programs that might otherwise be overlooked.

Typewriter

Christopher Latham Sholes, inventor, printer, and the editor of such newspapers as the *Milwaukee News* and the *Milwaukee Sentinel,* created the first modern typewriter and patented it in 1867. The earliest "writing machine" that was at all practicable had been devised by William Burt in 1829. The paper was fastened to a roller at the bottom of Burt's device, while the keys were on a circular carriage above it that had to be turned by hand to the proper location of each letter. Sholes invented the shift key, which made it possible to have both the lower case and capital letter on the same striking key, as well as the pianolike action of the keys, with the letters arrayed below the paper roller. In addition, his new carriage moved to the left one space at a time. He continued to perfect his machine for several years and then sold the rights to Remington, since he was unable to raise the money to start his own manufacturing company. Additional improvements were also given over to Remington.

In 1867, Charles E. Willer, a Milwaukee court reporter who was a friend of Sholes, came up with the famous test sentence to make sure the keys of a typewriter are functioning properly: "Now is the time for all good men to come to the aid of the party."

U

United Nations

The United Nations was not a purely American concept, but it bears an American stamp. In 1919, in the aftermath of World War I, President Woodrow Wilson conceived the idea for a League of Nations, and he received the Nobel Peace Prize in recognition of his efforts. But although many other countries joined the League, the

U.S. Congress, led by resurgent isolationist Republicans, refused to ratify the League and American participation in it. Without U.S. membership and with the rise of fascism in Europe and the belligerence of Japan in the Far East, the League foundered.

President Franklin Delano Roosevelt and Britain's Prime Minister Winston Churchill revived the idea of an international body to settle disputes, and they discussed it at length during Churchill's Christmas stay at the White House in 1941. It was Roosevelt who came up with the name United Nations. Eager to suggest the name to Churchill, Roosevelt arrived at the prime minister's door early on the morning of January 1, 1942, knocked, and was told to come in. A stark-naked Churchill, fresh from his bath, enthusiastically agreed to the name, noting the the phrase appeared in a poem by Lord Byron. A meeting of foreign ministers from twenty-six countries embraced the idea and the name that very day. The full UN charter was drawn up in San Francisco in September of 1945, and the organization's headquarters were established in New York City.

Unmanned Space Probe

The drama surrounding the Apollo program to land a man on the moon has tended to overshadow the great success of American unmanned space probes. The fact that the Soviet Union also had significant achievements in this area, sometimes achieving firsts before the U. S. did, also clouds the public perception of the American effort.

Some of the more notable accomplishments of NASA include:

Mariner 3 and *Mariner 4*, the first probes of Venus and Mars, respectively, in December 1962 and June 1965.

Surveyor 3, the first probe to soft land on the moon and scoop up lunar soil for testing, April 1967.

Mariner 9 was the first probe to orbit Mars, in November of 1971, sending back more than seven thousand pictures in the course of a year.

Pioneer 10, which sent back the first close-up views of Jupiter in December 1973, and went on to leave the solar system in 1986.

Viking 1, which landed on Mars in July of 1976 and sent back data for more than six years even though designed to last only ninety days.

Voyager 1 and *Voyager 2*, both of which passed Jupiter, Saturn, and Uranus sometime between 1979 and 1989.

Galileo, launched in 1989, will reach Jupiter in 1995, after a double "slingshot" voyage through the inner solar system.

V

Vichyssoise

Because of its name, most people think that this extremely popular cold soup was created in France. Although the chef who invented it was of French birth, he had been working in the United States for many years and was responding to American tastes when he concocted it.

Essentially he began with his mother's recipe for leek-and-potato soup. But that soup was always served hot and the chef changed the proportions and seasonings and puréed the blended ingredients before chilling them. Louis Diat, who headed the kitchen at New York's famed Ritz Carlton Hotel, first served vichysoisse in the summer of 1910. The soup has been a feature on American restaurant menus ever since, but it has never achieved great favor in France.

View From Space

One of the most wondrous side benefits of the Apollo space program were the many views sent back of our planet suspended in the firmament, a blue ball shining against the darkness of the vast reaches of the galaxy. But that view was topped in 1978, when *Voyager I,* on its way to Jupiter, sent back pictures of the earth and the moon together.

W

Walking Around the World

We are all encouraged to take a brisk walk as part of our daily exercise, but American David Kunst took that injunction rather seriously. He walked around the world. It took him from June 10, 1970, to October 5, 1974, a total of 1,576 days. He might have done it faster, but he took a route that allowed him to walk on dry land as much as possible and to spend the least time he could simply pacing the decks of a ship. It was this dedication to walking on actual land whenever feasible that gave validity to his claim of "walking" around the world.

No one has tried to duplicate the feat. It is the kind of thing that, done once, has well and truly been done for all time.

Walking on the Moon

In May of 1961, President John F. Kennedy issued one of the great challenges of human history: that the United States should land a man on the moon and return him safely to earth "before this decade is out." Beginning with the flight of *Apollo 8*, December 21–27, 1968, three flights were made to the moon and back before the go-ahead to land was scheduled for the flight of *Apollo 11*. The flight was commanded by a civilian, Neil A. Armstrong, joined by Air Force officers Col. Edwin E. "Buzz" Aldrin and Lt. Col. Michael Collins.

Apollo 11 took off on July 16, 1969, a Wednesday, and went into orbit around the moon on Saturday. At 3:08 P. M. eastern standard time, Armstrong and Aldrin, in the lunar module *Eagle*, detached from the command vessel *Columbia* and descended toward the surface of the moon. At 4:17 P. M., Aldrin's voice crackled back through space, "Houston, Tranquillity Base here. The *Eagle* has landed."

Three hours later, Neil Armstrong descended a nine-rung ladder to the lunar surface. As he did so, he opened the lens of a camera mounted in the module to record his historic first step onto the surface of the moon. On earth, more than five hundred million people watched him as he placed a foot in the lunar dust. "That's one small step for a man," he said, "one giant leap for mankind." One of the oldest dreams of humankind had at last been accomplished, and the first man on the moon was an American born in Wapakoneta, Ohio, in 1930.

War Photography

Mathew Brady was the most acclaimed portrait photographer of his time, from the 1840s to his death in 1896. From presidents John Quincy Adams, Andrew Jackson, John Tyler, James Polk, and Abraham Lincoln to an aged Dolley Madison and the world-famous singer Jenny Lind, Brady captured the likenesses of the great personalities of an entire age.

Handsome, wealthy, famous, he undertook to do something in 1861 that had never been done before—to photograph the course of a war at the front lines. "I felt I had to go," he said. "A spirit in my feet said 'Go' and I went . . ."

His monumental photographic coverage of the Civil War, carried out under the most difficult conditions, proved an epic achievement and one of the reasons that dreadful conflict remains so vivid to us even now. He photographed generals in dress uniform, ordinary soldiers eating, sleeping, cleaning their guns, and marching into battle. He caught the weary but stalwart expression of a male nurse—a little-known poet named Walt Whitman. He transcribed an experience that shapes America to this day and inspired generations of photographic artists who would also become war photographers.

Washington, D. C.

The choice of a capital city for the new United States was fraught with controversy. Boston, New York, Philadelphia, Williamsburg, and various other communities all had their adherents. As always, there was vast disagreement between the northern and southern states on the subject. As the founding fathers struggled to develop the original Constitution in 1887, a compromise was also worked out that called for a new "federal city," to be constructed from the ground up. Throughout history, even as the very shape and nature of countries and empires had changed continuously, capital cities had always been chosen from existing cities of importance. That the new country of the United States, its existence won through revolution, should choose to build a city where none had existed was a bold symbolic step. George Washington himself, originally a surveyor and mapmaker by profession, chose the site on the Potomac River where the city was to rise.

Wells, Fargo & Co.

Wells, Fargo & Co. was organized in 1852 by Henry Wells and William George Fargo as an express company to carry mail, gold dust, and passengers to California. Both men had previously been involved with an express operation between Albany and Buffalo, New York, and with the initial outing of the American Express Co. What was unique about Wells Fargo is that it ran not only an express but also banking facilities, a combination of services that enabled it to prosper against very tough competition. A number of mergers took place over the next two decades, and in 1873, all the operations were brought together under a reconstituted American Express Co., with Fargo as its first president.

The express activities of Wells Fargo should not be confused with the operation of the Pony Express, an American legand that was of neglible historical importance. Formed in 1860, it was in business for only a year, an early victim of the march of progress in the form of the transcontinental telegraph line that went into service in October 1861.

Woman's Solo Plane Flight

Kansas-born Amelia Mary Earhart was the first woman passenger on a transatlantic plane flight in 1928, but that was just the beginning of a remarkable career. In May 1932, Earhart flew the Atlantic solo in a record-breaking fourteen hours and fifty-six minutes. Not a lady to rest on her laurels, she made the first solo flight from Hawaii to the U. S. mainland in January 1935, and the first nonstop flight across the continent diagonally from Mexico City to Newark, New Jersey, in May of the same year. She also regularly broke the established speed records on several routes during the 1930s.

A major celebrity of her time, she embarked on a round-the-world flight in July of 1937, with navigator Frederick Noonan. Their plane disappeared between New Guinea and Hawkins Island in the South Pacific. Because she was regarded as one of the great pilots of history, the idea that she and her plane could simply vanish was taken as a sign that something mysterious had happened. There have been many attempts to locate the wreckage, and innumerable theories about her disappearance, ranging from the notion that she was caught in some kind of Pacific "Bermuda Triangle" to recent heavily documented investigations suggesting that she was deliberately shot down by the Japanese. Was she perhaps on a spy-plane mission for the U. S. government, attempting to take aerial photographs of Japanese installations in the Pacific? Experts think that may have been the case, but her disappearance remains a mystery unlikely to be solved.

Women's Magazine

The first national women's magazine, *Godey's Lady's Book* was cofounded in 1830 by Louis Antoine Godey and Sarah Josepha Hale, who edited the magazine for more than forty years. *Godey's* was dedicated to moral uplift, but it was cherished by women even more for its famous color illustrations of the latest fashions, elegant full-length drawings that are highly prized by collectors. Sarah Hale was a woman of many accomplishments. She led the campaign to make Thanksgiving a national holiday, and she wrote the immortal nursery rhyme "Mary Had a Little Lamb," which was based on an actual incident.

Woolworth's

Frank Winfield Woolworth was down on his luck when he moved to Lancaster, Pennsylvania, in 1879. A store he had owned in Utica, New York, had failed, and he was basically starting again. But he also had a concept: he would open a store carrying a wide variety of household goods to be sold at five or ten cents. This entirely new approach to marketing was a great success in Lancaster where one of the largest remaining Woolworth's in the country continues to operate. Opening new stores and buying up competitors, he rapidly built up the the largest chain-store empire in the county. Today's discount giants, Wal-Mart and Walgreen's, are direct descendants of Woolworth's original concept.

X

Xerox Machine

Photocopiers were invented as far back as 1938, by the physicist Chester Carlson. His revolutionary machine used no ink, but instead employed static electricity. A charged plate was suffused with light, which removed the electric charge from the white areas, while a plastic powder called toner was applied to the remaining areas, and the resulting printout reproduced the original.

Carlson called his new process *xerography*, from a combination of Greek words that literally mean "dry writing." It took another nine years to refine the process and bring down its cost to the point that it was commercially viable, and it was not until 1959, with the founding of the Xerox Corporation, that photocopying became a true wave of the future.

Z

Zipper

At the 1893 Chicago World's Fair one of the hundreds of inventions displayed a was patented new device called a "clasp-locker." A moveable hook-and-eye system, it was meant to be used on high boots, and its inventor, a mechanical engineer named Whitcomb Judson, wore a pair equipped with clasp-lockers himself. But the only interest in the device came from the U. S. Post Office, which ordered twenty test mail bags. The clasp-lockers kept jamming, however, and the bags were discarded. Judson continued to try to perfect the device, but there still seemed no market for it when he died in 1909.

A Swedish-American named Gideon Sundback, a former employee of Judson's, finally took another approach, and in 1914 Sundback perfected the zipper in the form we know today. The public still wasn't interested, but the U.S. Army was, and during World War I clothing and equipment for the armed forces began to feature zippers.

The next step was taken by the B. F. Goodrich company, which produced rubber boots with zippers starting in 1923. Mr. Goodrich himself coined the word *zipper*, which he felt mimicked the sound the device made when being opened and closed. As so often in American commerce, the right name attracted much greater interest, and by the late 1920s, concealed zippers were becoming common on many articles of clothing. Colored zippers were the brainchild of fashion designer Elsa Schiaparelli, and by the late 1930s, zippers had become a fashion statement in their own right, and have remained so to this day.

Bibliography

Ackerman, Diane. *A Natural History of the Senses.* New York: Random House, 1990.

Adler, Bill, Jr. *The Whole Earth Quiz Book.* New York: Morrow/Quill, 1991.

Berliner, Barbara, with Melinda Corey and George Ochoa. *The Book of Answers.* New York: Prentice Hall, 1990.

Brownstone, David, and Irene Franck. *20th Century Culture.* New York: Prentice Hall, 1991.

Clarke, Arthur C. *The Promise of Space.* New York: Harper & Row, 1968.

Clayman, Charles B., ed. *The American Medical Association Home Medical Encyclopedia,* vols. 1 & 2. New York: Random House, 1989.

Concise Dictionary of American Biography. New York: Scribners, 1964.

Davis, Mac. *Giant Book of Sports.* New York: Grosset & Dunlap, 1967.

Debono, G., ed. *Eureka! The History of Invention.* New York: Holt, Rinehart & Winston, 1974.

Douglas, Harvey, and Don E. Fehrenbacher, eds. *The Illustrated Biographical Dictionary.* New York: Dorset Press, 1990.

Flexner, Stuart Berg. *I Hear America Singing.* New York: Van Nostrand Reinhold, 1976.

Fuld, James. *The Book of World Famous Music.* Mineola, N. Y.: Dover, 1985.

Grun, Bernard. *The Timetables of History.* New York: Simon and Schuster (Touchstone), 1982.

Guinness Book of World Records. New York: Bantam Books, 1990.

Josephy, Alvin M., Jr. *The Indian Heritage of America.* New York: Alfred A. Knopf, 1968.

Katz, Ephraim. *The Film Encyclopedia.* New York: Perigee Books, 1979.

Levey, Judith S., and Agnes Greenhall. *The Concise Columbia Encyclopedia.* New York: Avon Books, 1983.

McGee, Harold. *On Food and Cooking.* New York: Scribners, 1984.

McGrath, Molly Wade. *Topsellers.* New York: Morrow, 1983.

Malone, Dumas, and Basil Rauch. *Empire for Liberty.* New York: Appleton-Century Croft, 1960.

Meserole, Mike, ed. *The 1992 Information Please Sports Almanac.* Boston: Houghton Mifflin, 1991.

150 Years of Baseball. Lincolnwood, Ill.: Publications International, 1989.

O'Neill, Molly. *New York Cookbook.* New York: Workman, 1992.

Panati, Charles. *Panati's Extraordinary Origins of Everyday Things.* New York: Harper and Row, 1991.

———*Panati's Parade of Fads, Follies and Manias.* New York: Harper and Row, 1987.

Reader's Digest How in the World. New York: The Reader's Digest Association, 1990.

The Smithsonian Book of Invention. New York: W. W. Norton, 1978.

Tuleja, Tad. *Fabulous Fallacies.* New York: Harmony Books, 1982.

Visser, Margaret. *Much Depends on Dinner.* New York: Grove Press, 1986.

Wiley, Mason, and Damien Bona. *Inside Oscar.* New York: Ballantine Books, 1986.

Wilson, Mitchell. *American Science and Invention.* New York: Bonanza Books, 1960.

ABOUT THE AUTHOR

NORMAN KING, international investment banker, lectures widely and is the author of sixteen books. His fascination with celebrities and the business of show business has resulted in biographies of Dan Rather, Erma Bombeck, Oprah Winfrey, Pricesses Di and Fergie, Madonna, Phil Donahue, Arsenio Hall, and Hillary Clinton.

Educated at Cornell, New York University, and the New York University Law School, King resides in New York City with his wife, Barbara.

More Fun Facts and Interesting Trivia

Ask for any of the books listed below at your bookstore. Or to order direct from the publisher, call 1-800-447-BOOK (MasterCard or Visa), or send a check or money order for the books purchased (plus $4.00 shipping and handling for the first book ordered and 75¢ for each additional book) to Carol Publishing Group, 120 Enterprise Avenue, Dept. 1549, Secaucus, NJ 07094.

Aardvarks to Zebras: A Menagerie of Facts, Fiction and Fantasy About the Wonderful World of Animals by Melissa S. Tulin
Illustrated with photographs & drawings
$14.95 paper   0-8065-1548-1   (CAN $20.95)

The Almanac of Fascinating Beginnings: From the Academy Awards to the Xerox Machine by Norman King
$9.95 paper   0-8065-1549-X   (CAN $13.95)

The Book of Totally Useless Information by Donald A. Voorhees
$7.95 paper   0-8065-1405-1   (CAN $9.95)

The Business Disasters Book of Days: The World's Greatest Financial Mishaps, Follies and Remarkable Events by Jill Herbers
$9.95 paper   0-8065-1585-6   (CAN $13.95)

The "Cheers" Trivia Book by Mark Wenger
Illustrated with photographs throughout
$9.95 paper   0-8065-1482-5   (CAN $11.95)

The Complete Book of Sexual Trivia by Leslee Welch
$7.95 paper   0-8065-1347-0   (CAN $9.95)

Death by Rock & Roll: The Untimely Deaths of the Legends of Rock by Gary J. Katz
Illustrated with photographs throughout
$9.95 paper   0-8065-1581-3   (CAN $13.95)

The Encyclopedia of Popular Misconceptions: The Ultimate Debunker's Guide to Widely Accepted Fallacies by Ferris Johnsen
$9.95 paper   0-8065-1556-2   (CAN $13.95)

50 Greatest Conspiracies of All Time: History's Biggest Mysteries, Cover-ups and Cabals by Jonathan Vankin and John Whalen
$12.95 paper   0-8065-1576-7   (CAN $17.95)

Film Flubs: Memorable Movie Mistakes by Bill Givens   Illustrated with photographs throughout
$7.95 paper   0-8065-1161-3   (CAN $10.95)
Also available:
Son of Film Flubs: More Memorable Movie Mistakes by Bill Givens
Illustrated with photographs throughout
$7.95 paper   0-8065-1279-2   (CAN $10.95)

Film Flubs: The Sequel - Even More Memorable Movie Mistakes by Bill Givens
Illustrated with photographs throughout
$7.95 paper   0-8065-1360-8   (CAN $9.95)

George Washington Had No Middle Name: Strange Historical Facts by Patricia Lee Holt
$5.95 paper   0-8065-1074-9   (CAN $7.95)

How a Fly Walks Upside Down and Other Curious Facts by Martin A. Goldwyn
Illustrations throughout
$7.95 paper   0-8065-1054-4   (CAN $10.95)

Light Your House With Potatoes: And 99 Other Off-the-wall Solutions to Life's Little Problems by Jay Kaye   Illustrations throughout
$7.95 paper   0-8065-1376-4   (CAN $9.95)

Old Wives' Tales: The Truth Behind Common Notions by Sue Castle   Illustrations throughout
$7.95 paper   0-8065-1378-0   (CAN $9.95)

1001 Toughest TV Trivia Questions of All Time: Brain Teasers for the Thinking Couch Potato by Vincent Terrace
$9.95 paper   0-8065-1499-X   (CAN $11.95)

Peculiar Patents: A Collection of Unusual and Interesting Inventions from the Files of the U.S. Patent Office by Dr. Rick Feinberg
Illustrated throughout
$9.95 paper   0-8065-1561-9   (CAN $13.95)

The "Seinfeld" Aptitude Test: Hundreds of Spectacular Questions on Minute Details from TV's Greatest Show About Absolutely Nothing by Beth B. Golub
$8.95 paper   0-8065-1583-X   (CAN $12.95)

Sophomore Slumps: Disastrous Second Movies, Albums, Songs and TV Shows by Chris Golden   Illustrated with photographs throughout
$9.95 paper   0-8065-1584-8   (CAN $13.95)

Why Do They Call It Topeka?: How Places Got Their Names by John W. Pursell
$9.95 paper   0-8065-1588-0   (CAN $13.95)

(Prices subject to change; books subject to availability)